Library of
Davidson College

body movement
PERSPECTIVES IN RESEARCH

body movement
PERSPECTIVES IN RESEARCH

Advisory Editor: Martha Davis
Hunter College

ANTHROPOLOGICAL PERSPECTIVES OF MOVEMENT

ARNO PRESS
A NEW YORK TIMES COMPANY
New York • 1975

Reprint Edition 1975 by Arno Press Inc.

Copyright © 1975, by Arno Press Inc.

<u>The Cultural Basis of Emotions and Gestures</u>   1947 by Duke University Press, Durham, North Carolina, reprinted by permission of Duke University Press and Weston LaBarre

<u>Panorama of Dance Ethnology</u>   1960 by the Wenner-Gren Foundation for Anthropological Research, Inc., reprinted by permission of The University of Chicago Press and Dr. Gertrude Kurath

Body Movement: Perspectives in Research
ISBN for complete set: 0-406-06197-8
See last pages of this volume for titles.

Manufactured in the United States of America

Library of Congress Cataloging in Publication Data
Main entry under title:

Anthropological perspectives of movement.

   (Body movement: perspectives in research)
   1. Man--Attitude and movement--Addresses, essays, lectures. 2. Gesture--Addresses, essays, lectures. I. Series. [DNLM: 1. Anthropology, Physical. 2. Movement. GN231 A628]
GN231.A57   301.2'1   74-9162
ISBN 0-405-06201-X

Contents

LaBarre, Weston
THE CULTURAL BASIS OF EMOTIONS AND GESTURES
(*Reprinted from* JOURNAL OF PERSONALITY, Vol. 16, 1947)

Bailey, Flora L.
NAVAHO MOTOR HABITS (*Reprinted from* AMERICAN ANTHROPOLOGIST, Vol. 44, 1942)

Hewes, Gordon W.
WORLD DISTRIBUTION OF CERTAIN POSTURAL HABITS
(*Reprinted from* AMERICAN ANTHROPOLOGIST, Vol. 57, 1955)

Kurath, Gertrude Prokosch
PANORAMA OF DANCE ETHNOLOGY (*Reprinted from* CURRENT ANTHROPOLOGY, Vol. 1, May 1960)

Hall, Edward T.
A SYSTEM FOR THE NOTATION OF PROXEMIC BEHAVIOR
(*Reprinted from* AMERICAN ANTHROPOLOGIST, Vol. 65, 1963)

Hayes, Francis
GESTURES: A Working Bibliography (*Reprinted from* SOUTHERN FOLKLORE QUARTERLY, Vol. 21, 1957)

THE CULTURAL BASIS OF EMOTIONS AND GESTURES

WESTON LaBARRE

THE CULTURAL BASIS OF EMOTIONS AND GESTURES

WESTON LABARRE
Duke University

Psychologists have long concerned themselves with the physiological problems of emotion, as for example, whether the psychic state is prior to the physiological changes and causes them, or whether the conscious perception of the inner physiological changes in itself constitutes the "emotion." The physiologists also, notably Cannon, have described the various bodily concomitants of fear, pain, rage, and the like. Not much attention, however, has been directed toward another potential dimension of meaning in the field of emotions, that is to say the *cultural* dimension.

The anthropologist is wary of those who speak of an "instinctive" gesture on the part of a human being. One important reason is that a sensitivity to meanings which are culturally different from his own stereotypes may on occasion be crucial for the anthropologist's own physical survival among at least some groups he studies, and he must at the very least be a student of this area of symbolism if he would avoid embarrassment.[1] He cannot safely rely upon his own culturally subjective understandings of emotional expression in his relations with persons of another tribe. The advisability and the value of a correct reading of any cultural symbolism whatsoever has alerted him to the possibility of culturally arbitrary, quasi-linguistic (that is, noninstinctual but learned and purely agreed-upon) meanings in the behavior he observes.

A rocking of the skull forward and backward upon its condyles, which rest on the atlas vertebra, as an indication of affirmation and the rotation upon the axis vertebra for negation have so far been accepted as "natural" and "instinctive" gestures that one psychol-

---

[1] The notorious Massey murder in Hawaii arose from the fact that a native beach boy perhaps understandably mistook the Occidental "flirting" of a white woman for a *bona fide* sexual invitation. On the other hand, there are known cases which have ended in the death of American ethnographers who misread the cultural signs while in the field.

ogist at least[2] has sought an explanation of the supposedly universal phenomenon in ascribing the motions of "yes" to the infant's seeking of the mother's breast, and "no" to its avoidance and refusal of the breast. This is ingenious, but it is arguing without one's host, since the phenomenon to be explained is by no means as widespread ethnologically, even among humans, as is mammalian behavior biologically.

Indeed, the Orient alone is rich in alternatives. Among the Ainu of northern Japan, for example, our particular head noddings are unknown:

the right hand is usually used in negation, passing from right to left and back in front of the chest; and both hands are gracefully brought up to the chest and gracefully waved downwards—palms upwards—in sign of affirmation.[3]

The Semang, pygmy Negroes of interior Malaya, thrust the head sharply forward for "yes" and cast the eyes down for "no."[4]

The Abyssinians say "no" by jerking the head to the right shoulder, and "yes" by throwing the head back and raising the eyebrows. The Dyaks of Borneo raise their eyebrows to mean "yes" and contract them slightly to mean "no." The Maori say "yes" by raising the head and chin; the Sicilians say "no" in exactly the same manner.[5]

A Bengali servant in Calcutta rocks his head rapidly in an arc from shoulder to shoulder, usually four times, in assent; in Delhi a Moslem boy throws his head diagonally backward with a slight turning

---

[2] E. B. Holt, *Animal drive and the learning process* (New York, 1931), p. 111, and personal conversations.

The idea is originally Darwin's, I believe (Charles Darwin, *The expression of the emotions in man and animals*, New York, 1873), but he himself pointed out that the lateral shake of the head is by no means universally the sign of negation. Holt has further noted the interesting point that in a surprising number of languages, quite unrelated to each other, the word for "mother" is a variant of the sound "ma." One can collect dozens of such instances, representing all the continents, which would seem to confirm his conjecture: the genuinely universal "sucking reflex" which brings the lips into approximation (m), plus the simplest of the simple open vowel sounds (a), are "recognized" by the mother as referring to her when the baby first pronounces them; hence they become the lexical designation of the maternal parent. Although this phenomenon becomes a linguistic one, it is only on some such physiological basis that one can explain the recurrence of the identical sound combinations in wholly unrelated languages referring to the same person, the mother. But there is no absolute semantic association involved: one baby boy I have observed used "mama" both to connote and to denote older persons of either sex.

[3] A. H. S. Landor, *Alone with the Hairy Ainu* (London, 1893), pp. 6, 233-234.

[4] W. W. Skeat and C. O. Blagden, *Pagan races of the Malay Peninsula* (London, 1906; 2 vols.).

[5] Otto Klineberg, *Race differences* (New York, 1935), p. 282.

of the neck for the same purpose and the Kandyan Singhalese bends the head diagonally forward toward the right, with an indescribably graceful turning in of the chin, often accompanying this with a cross-legged curtsey, arms partly crossed, palms upward—the whole performance extraordinarily beautiful and ingratiating. Indeed, did my own cultural difference not tell me it already, I would know that the Singhalese manner of receiving an object (with the right hand, the left palm supporting the right elbow) is not instinctive, for I have seen a Singhalese mother *teaching* her little boy to do this when I gave him a chunk of palm-tree sugar. I only regretted, later, that my own manners must have seemed boorish or subhuman, since I handed it to him with my right hand, instead of with both, as would any courteous Singhalese. Alas, if I had handed it to a little Moslem beggar in Sind or the Punjab with my *left* hand, he would probably have dashed the gift to the ground, spat, and called me by the name of an animal whose flesh he had been taught to dislike, but which I have not—for such use of the left hand would be insulting, since it is supposed to be confined to attending to personal functions, while the right hand is the only proper one for food.

Those persons with a passion for easy dominance, the professional dog-lovers, must often be exasperated at the stupidity of a dog which does not respond to so obvious a command as the pointed forefinger. The defense of man's best friend might be that this "instinctively" human gesture does not correspond to the kinaesthesias of a nonhanded animal. Nevertheless, even for an intelligent human baby, at the exact period when he is busily using the forefinger in exploring the world, "pointing" by an adult is an arbitrary, sublinguistic gesture which is not automatically understood and which must be *taught*. I am the less inclined to berate the obtuseness to the obvious of either dog or baby, because of an early field experience of my own. One day I asked a favorite informant of mine among the Kiowa, old Mary Buffalo, where something was in the *ramada* or willow-branch "shade" where we were working. It was clear she had heard me, for her eighty-eight-year-old ears were by no means deaf; but she kept on busying both hands with her work. I wondered at her rudeness and repeated the request several times, until finally with a puzzled exasperation which matched my own, she dropped her work and fetched it for me from in plain sight: she had been repeatedly pointing with her lips in approved American Indian fashion, as any Caucasian numbskull should have been able to see.

Some time afterward I asked a somewhat naïve question of a very great anthropologist, the late Edward Sapir: "Do other tribes cry and laugh as we do?" In appropriate response, Sapir himself laughed, but with an instant grasping of the point of the question: In which of these things are men alike everywhere, in which different? Where are the international boundaries between physiology and culture? What are the extremes of variability, and what are the scope and range of cultural differences in emotional and gestural expression? Probably one of the most learned linguists who have ever lived, Sapir was extremely sensitive to emotional and sublinguistic gesture—an area of deep illiteracy for most "Anglo-Saxon" Americans—and my present interest was founded on our conversation at that time.

Smiling, indeed, I have found may almost be mapped after the fashion of any other culture trait; and laughter is in some senses a geographic variable. On a map of the Southwest Pacific one could perhaps even draw lines between areas of "Papuan hilarity" and others where a Dobuan, Melanesian dourness reigned. In Africa, Gorer noted that

> laughter is used by the negro to express surprise, wonder, embarrassment and even discomfiture; it is not necessarily, or even often a sign of amusement; the significance given to "black laughter" is due to a mistake of supposing that similar symbols have identical meanings.[6]

Thus it is that even if the physiological behavior be present, its cultural and emotional functions may differ. Indeed, even within the same culture, the laughter of adolescent girls and the laughter of corporation presidents can be functionally different things; so too the laughter of an American Negro and that of the white he addresses.

The behaviorist Holt "physiologized" the smile as being ontogenetically the relaxation of the muscles of the face in a baby replete from nursing. Explanations of this order may well be the case, if the phenomenon of the smile is truly a physiological expression of generalized pleasure, which is caught up later in ever more complex

---

[6] Geoffrey Gorer, *Africa dances* (New York, 1935), p. 10.

conditioned reflexes. And yet, even in its basis here, I am not sure that this is the whole story: for the "smile" of a child in its sleep is certainly in at least some cases the grimace of *pain* from colic, rather than the relaxation of pleasure. Other explanations such as that the smile is *phylogenetically* a snarl suffer from much the same *ad hoc* quality.

Klineberg writes:

> It is quite possible, however, that a smile or a laugh may have a different meaning for groups other than our own. Lafcadio Hearn has remarked that the Japanese smile is not necessarily a spontaneous expression of amusement, but a law of etiquette, elaborated and cultivated from early times. It is a silent language, often seemingly inexplicable to Europeans, and it may arouse violent anger in them as a consequence. The Japanese child is taught to smile as a social duty, just as he is taught to bow or prostrate himself; he must always show an appearance of happiness to avoid inflicting his sorrow upon his friends. The story is told of a woman servant who smilingly asked her mistress if she might go to her husband's funeral. Later she returned with his ashes in a vase and said, actually laughing, "Here is my husband." Her White mistress regarded her as a cynical creature; Hearn suggests that this may have been pure heroism.[7]

Many in fact of these motor habits in one culture are open to grave misunderstanding in another. The Copper Eskimo welcome strangers with a buffet on the head or shoulders with the fist, while the northwest Amazonians slap one another on the back in greeting. Polynesian men greet each other by embracing and rubbing each other's back; Spanish-American males greet one another by a stereotyped embrace, head over right shoulder of the partner, three pats on the back, head over reciprocal left shoulder, three more pats. In the Torres Straits islands "the old form of greeting was to bend slightly the fingers of the right hand, hook them with those of the person greeted, and then draw them away so as to scratch the palm of the hand; this is repeated several times."[8] The Ainu of Yezo have a peculiar greeting; on the occasion of a man meeting his sister, "The man held the woman's hands for a few seconds, then suddenly releasing his hold, grasped her by both ears and uttered the Aino cry. Then they stroked one another down the face and

---

[7] Lafcadio Hearn, The Japanese smile, in *Glimpses of unfamiliar Japan* (New York, 1894; 2 vols.), quoted in Klineberg, *op. cit.*
[8] *Report on the Cambridge expedition to the Torres Straits*, ed. A. C. Haddon (Cambridge, 1904; 5 vols.), IV, p. 306; Thomas Whiffen, *The North West Amazons* (London, 1905), p. 259.

shoulders."[9] Kayan males in Borneo embrace or grasp each other by the forearm, while a host throws his arm over the shoulder of a guest and strokes him endearingly with the palm of his hand. When two Kurd males meet, "they grasp each other's right hand, which they simultaneously raise, and each kisses the hand of the other."[10] Among the Andaman Islanders of the Gulf of Bengal:

When two friends or relatives meet who have been separated from each other for a few weeks or longer, they greet each other by sitting down, one on the lap of the other, with their arms around each other's necks, and weeping or wailing for two or three minutes till they are tired. Two brothers greet each other in this way, and so do father and son, mother and daughter, and husband and wife. When husband and wife meet, it is the man who sits in the lap of the woman. When two friends part from one another, one of them lifts up the hand of the other towards his mouth and gently blows on it.[11]

Some of these expressions of "joy" seem more lugubrious than otherwise. One old voyager, John Turnbull, writes as follows:

The arrival of a ship brings them to the scene of action from far and near. Many of them meet at Matavai who have not seen each other for some length of time. The ceremony of these meetings is not without singularity; taking a shark's tooth, they strike it into their head and temples with great violence, so as to produce a copious bleeding; and this they will repeat, till they become clotted with blood and gore.

The honest mariner confesses to be nonplussed at this behavior.

I cannot explain the origin of this custom, nor its analogy with what it is intended to express. It has no other meaning with them than to express the excess of their joy. By what construction it is considered symbolical of this emotion I do not understand.[12]

Quite possibly, then, the weeping of an American woman "because she is so happy" may merely indicate that the poverty of our gamut of physiological responses is such as to require using the same response for opposite meanings. Certainly weeping does obey social stereotypes in other cultures. Consider old Mary Buffalo at her

[9] R. Hitchcock, The Ainos of Yezo, in *Papers on Japan*, pp. 464-465. See also Landor, *op. cit.*, pp. 6, 233-234.
[10] J. Perkins, Journal of a tour from Oroomish to Mosul, through the Koordish Mountains, and a visit to the ruins of Nineveh, *Journal of the American Oriental Society*, 1851, 2, 101; Charles Hose & William MacDougall, *The pagan tribes of Borneo* (London, 1912; 2 vols.), I, 124-125.
[11] A. R. Radcliffe-Brown, *The Andaman Islanders* (Cambridge, 1922), p. 117 and p. 74 n. 1.
[12] John Turnbull, *A voyage round the world* (London, 1813), pp. 301-302.

brother's funeral: she wept in a frenzy, tore her hair, scratched her cheeks, and even tried to jump into the grave (being conveniently restrained from this by remoter relatives). I happened to know that she had not seen her brother for some time, and there was no particular love lost between them: she was merely carrying on the way a decent woman should among the Kiowa. Away from the grave, she was immediately chatting vivaciously about some other topic. Weeping is *used* differently among the Kiowa. Any stereotypes I may have had about the strong and silent American Indian, whose speech is limited to an infrequent "ugh" and whose stoicism to pain is limitless, were once rudely shattered in a public religious meeting. A great burly Wichita Indian who had come with me to a peyote meeting, after a word with the leader which I did not understand (it was probably permission to take his turn in a prayer) suddenly burst out blubbering with an abandon which no Occidental male adult would permit himself in public. In time I learned that this was a stereotyped approach to the supernatural powers, enthusiastic weeping to indicate that he was as powerless as a child, to invoke their pity, and to beseech their gift of medicine power. Everyone in the tipi understood this except me.

So much for the expression of emotion in one culture, which is open to serious misinterpretation in another: there is no "natural" language of emotional gesture. To return a moment to the earlier topic of emotional expression in greetings: West Africans in particular have developed highly the ritual gestures and language of greeting. What Gorer says of the Wolof would stand for many another tribe:

The gestures and language of polite intercourse are stylized and graceful; a greeting is a formal litany of question and answer embracing everyone and everything connected with the two people meeting (the questions are merely formal and a dying person is stated to be in good health so as not to break the rhythm of the responses) and continuing for several minutes; women accompany it with a swaying movement of the body; with people to whom a special deference is due the formula is resumed several times during the conversation; saying goodbye is equally elaborate.[13]

[13] Gorer, *op. cit.*, p. 38. Cf. Hollis, *The Masai, their language and folklore* (Oxford, 1905), pp. 284-287; E. Torday & T. A. Joyce, *Notes ethnographiques sur les peuples communément appelés Bakuba, ainsi que sur les peuplades apparentées, les Bushonga* (Brussels, 1910), pp. 233-234, 284, *et passim*. West Africans have developed the etiquette and protocol of greeting to a high degree, adjusting it to sex, age, relative rank, relationship degrees, and the like. Probably there is more than a trace of this ceremoniousness surviving in American Negro greetings in the South.

But here the sublinguistic gesture language has clearly emerged into pure formalisms of language which are quite plainly cultural.

The allegedly "instinctive" nature of such motor habits in personal relationships is difficult to maintain in the face of the fact that in many cases the same gesture means exactly opposite, or incommensurable things, in different cultures. Hissing in Japan is a polite deference to social superiors; the Basuto applaud by hissing, but in England hissing is rude and public disapprobation of an actor or a speaker. Spitting in very many parts of the world is a sign of utmost contempt; and yet among the Masai of Africa it is a sign of affection and benediction, while the spitting of an American Indian medicine man upon a patient is one of the kindly offices of the curer. Urination upon another (as in a famous case at the Sands Point, Long Island, country club, involving a congressman since assassinated) is a grave insult among Occidentals, but it is part of the transfer of power from an African medicine man in initiations and curing rituals. As for other opposite meanings, Western man stands up in the presence of a superior; the Fijians and the Tongans sit down. In some contexts we put on more clothes as a sign of respect; the Friendly Islanders take them off. The Toda of South India raise the open right hand to the face, with the thumb on the bridge of the nose, to express respect; a gesture almost identical among Europeans is an obscene expression of extreme disrespect. Placing to the tip of the nose the projecting knuckle of the right forefinger bent at the second joint was among the Maori of New Zealand a sign of friendship and often of protection;[14] but in eighteenth-century England the placing of the same forefinger to the right side of the nose expressed dubiousness about the intelligence and sanity of a speaker—much as does the twentieth-century clockwise motion of the forefinger above the right hemisphere of the head. The sticking out of the tongue among Europeans (often at the same time "making a face") is an insulting, almost obscene act of provocative challenge and mocking contempt for the adversary, so undignified as to be used only by children; so long as Maya writing remains undeciphered we do not know the meaning of the exposure of the

[14] Klineberg, *op. cit.*, pp. 286-287, citing J. Lubbock, *Prehistoric times* (New York, 1872); E. Best, *The Maori* (Wellington [N. Z.], 1924; 2 vols.); R. H. Lowie, *Are we civilized?* (New York, 1929); and A. C. Hollis, *The Masai, their language and folklore.* (Oxford, 1905), p. 315.

tongue in some religious sculptures of the gods, but we can be sure it scarcely has the same significance as with us. In Bengali statues of the dread black mother goddess Kali, the tongue is protruded to signify great raging anger and shock; but the Chinese of the Sung dynasty protruded the tongue playfully to pretend to mock terror, as if to "make fun of" the ridiculous and unfeared anger of another person.[15] Modern Chinese, in South China at least, protrude the tongue for a moment and then retract it, to express embarrassment at a *faux pas*.

Kissing, as is well known, is in the Orient an act of private love-play and arouses only disgust when indulged in publicly: in Japan it is necessary to censor out the major portion of love scenes in American-made movies for this reason. Correspondingly, some of the old *kagura* dances of the Japanese strike Occidentals as revoltingly overt obscenities, yet it is doubtful if they arouse this response in Japanese onlookers. Manchu kissing is purely a private sexual act, and though husband and wife or lovers might kiss each other, they would do it stealthily since it is shameful to do in public; yet Manchu mothers have the pattern of putting the penis of the baby boy into their mouths, a practice which probably shocks Westerners even more than kissing in public shocks the Manchu.[16] Tapuya men in South America kiss as a sign of peace, but men do not kiss women because the latter wear labrets or lip plugs. Nose-rubbing is Eskimo and Polynesian; and the Djuka Negroes of Surinam[17] show pleasure at a particularly interesting or amusing dance step by embracing the dancer and touching cheek to cheek, now on one side, now on the other—which is the identical attenuation of the "social kiss" between American women who do not wish to spoil each other's makeup.

In the language of gesture all over the world there are varying mixtures of the physiologically conditioned response and the purely cultural one, and it is frequently difficult to analyze out and segregate the two. The Chukchee of Siberia, for example, have a phe-

[15] *Chin P'ing Mei* (Shanghai, n. d.), Introduction by Arthur Waley. The sixteenth-century Chinese also had the expressions to act "with seven hands and eight feet" for awkwardness, and "to sweat two handfuls of anxiety."
[16] S. M. Shirokogoreff, *Social organization of the Manchus* (Extra Vol. III, North China Branch, Royal Asiatic Society, Shanghai, 1924), pp. 122-123.
[17] M. C. Kahn, Notes on the Saramaccaner Bush Negroes of Dutch Guiana. *Amer. Anthrop.*, 1929, **31**, 473.

nomenal quickness to anger, which they express by showing the teeth and growling like an animal—yet man's snout has long ceased being functionally useful in offensive or defensive biting as it has phylogenetically and continuously retreated from effective prognathism. But this behavior reappears again and again: the Malayan pagans, for example, raise the lip over the canine tooth when sneering and jeering. Is this instinctual reflex or mere motor habit? The Tasmanians stamped rapidly on the ground to express surprise or pleasure; Occidentals beat the palms of the hands together for the same purpose ordinarily, but in some rowdier contexts this is accompanied by whistling and a similar stamping of the feet. Europeans "snort" with contempt; and the non-Mohammedan primitives of interior Malaya express disgust with a sudden expiration of the breath. In this particular instance, it is difficult to rid oneself of the notion that this is a consciously controlled act, to be sure, but nevertheless at least a "symbolic sneeze" based upon a purely physiological reflex which does rid the nostrils of irritating matter. The favorite gesture of contempt of the Menomini Indians of Wisconsin—raising the clenched fist palm down up to the level of the mouth, then bringing it swiftly downwards, throwing forth the thumb and first two fingers—would seem to be based on the same "instinctual" notion of rejection.

However, American Indian gestures soon pass over into the undisputedly linguistic area, as when two old men of different tribes who do not know a word of each other's spoken language, sit side by side and tell each other improper stories in the complex and highly articulate intertribal sign language of the Plains. These conventionalized gestures of the Plains sign language must of course be learned as a language is learned, for they are a kind of kinaesthetic ideograph, resembling written Chinese. The written Chinese may be "read" in the Japanese and the Korean and any number of mutually unintelligible spoken Chinese dialects; similarly, the sign language may be "read" in Comanche, in Cheyenne, or in Pawnee, all of which belong to different language families. The primitive Australian sign language was evidently of the conventionalized Plains type also, for it reproduced words, not mere letters (since of course they had no written language), but unfortunately little is known in detail of its mechanisms.

Like the writing of the Chinese, Occidental man has a number of ideographs, but they are sublinguistic and primarily *signs to action* or *expressions of action*. Thus, in the standard symbolism of cartoons, a "balloon" encircling print has signified *speaking* since at least the eighteenth century. Interestingly, in a Maya painting on a vase from Guatemala of pre-Columbian times, we have the same speech "balloons" enclosing ideographs representing what a chief and his vassal are saying, though what that is we do not know.[18] In Toltec frescoes speech is symbolized by foliated or noded crooks or scrolls, sometimes double, proceeding out of the mouths of human figures, although *what* is said is not indicated.[19] In the later Aztec codices written on wild fig-bark paper, speech is conventionalized by one or more little scrolls like miniature curled ostrich feathers coming out of the mouths of human beings, while motion or walking is indicated by footprints leading to where the person is now standing in the picture.[20] In American cartoons the same simple idea of footprints is also used. The ideograph of "sawing wood" indicates the action of *snoring* or *sleeping*. A light bulb with radial lines means that a "bright idea" has just occurred in the mind of the character above whose head it is written. While even children learn in time to understand these signs in context, no one would maintain that the electric-light "sign" could naturally be understood by an individual from another tribe than our own. Birds singing, a spiral, or a five-pointed star means unconsciousness or semiconsciousness through concussion. A dotted line, if curved, indicates the past trajectory of a moving object; if straight and from eye to object, the action of seeing. None of these visual aids to understanding are part of objective nature. Sweat drops symbolize surprise or dumfounding, although the physiology of this sign is thoroughly implausible. And]%!*/=#?[¢& very often says the unspeakable, quite as ? signifies query and ! surprise.

Many languages have *spoken* punctuation marks, which English grievously lacks. On the other hand, the speakers of English have a few *phonetic* "ideographs," at least two of which invite to action. An imitation of a kiss, loudly performed, summons a dog, if that

[18] George C. Vaillant, *The Aztecs of Mexico* (Garden City, N. Y., 1941), plate 7, top.
[19] *Ibid.*, plate 24.
[20] *Ibid.*, plates 42, 57, 61.

dog understands this much of English. A bilateral clucking of the tongue adjures a horse to "giddyap," i.e., to commence moving or to move more smartly; and in some parts of the country at least, it has a secondary semantic employment in summoning barnyard fowl to their feeding. The dental-alveolar repeated clicking of the tongue, on the other hand, is not a symbolic ideophone to action, but a *moral comment* upon action, a strongly critical disapprobation largely confined in use to elderly females preoccupied with such moral commentary. These symbolic ideophones are used in no other way in our language; but in African Bushman and Hottentot languages, of course, these three sounds plus two others phonetically classified as "clicks" (as opposed to sonants like b, d, g, z and surds like p, t, k, s, etc.) are regularly employed in words like any other consonants. It is nonsense to suppose that dogs, horses, or chickens are equipped for "instinctive" understanding or response to these human-made sounds, as much as that speakers of English have an instinctive understanding of Hottentot and Bushman. Certainly the sounds used in the Lake Titicaca plateau to handle llamas are entirely different.[21]

Sublinguistic "language" can take a number of related forms. Among the Neolithic population of the Canary Islands there was a curious auxiliary "language" of conventionalized whistles, signals which could be understood at greater distances than mere spoken speech. On four bugle tones, differently configurated, we can similarly order soldiers to such various actions as arising, assembling, eating, lowering a flag, and burying the dead. The drum language of West Africa, however, is more strictly linguistic than bugle calls. Many West African languages are tonemic, that is, they have pitch-accent somewhat like Chinese or Navaho. Drum language, therefore, by reproducing not only the rhythm but also the tonal configurations of familiar phrases and sentences, is able to send messages of high semantic sophistication and complexity, as easily recognizable as our "Star-Spangled Banner" sung with rhythm and melody, but without words. The Kru send battle signals on

[21] Weston La Barre, *The Aymara Indians of the Lake Titicaca Plateau,* Memoir 68, Amer. Anthrop. Assn. (Menasha, Wisconsin, 1947). All the tribes of the Provincia Oriental of Ecuador had the "cluck of satisfaction" (Alfred Simpson, *Travels in the wilds of Ecuador and exploration of the Putumayo River,* 1886, p. 94), which among the tribes of the Issa-Japura rivers is a "sign of assent and pleasure" (Whiffen, *op. cit.,* p. 249).

multiple-pitched horns, but these are not conventional tunes like our bugle calls, but fully articulated sentences and phrases whose tonemic patterns they reproduce on an instrument other than the human vocal cords. The Morse and International telegraph codes and Boy Scout and Navy flag communication (either with hand semaphores or with strings of variously shaped and colored flags) are of course mere auditory or visual alphabets, tied down except for very minor conventionalized abbreviations to the *spelling* of a given language. (The advantages of a phonetic script, however, are very evident when it comes to sending messages via a Morselike code for Chinese, which is written in ideographs which have different phonetic pronunciations in different dialects; Japanese has some advantage in this situation over Chinese in that its ideographs are already cumbrously paralleled in *katakana* and *hiragana* writing, which is quasi-phonetic.) Deaf-and-dumb language, if it is the mere spelling of words, is similarly bound to an alphabet; but as it becomes highly conventionalized it approaches the international supralinguistic nature of the Plains Indian sign language. Of this order are the symbols of mathematics, the conventionalizations on maps for topography, the symbol language for expressing meteorological happenings on weather maps, and international flag signals for weather. Modern musical notation is similarly international: a supralinguistic system which orders in great detail what to do, and with what intensity, rhythm, tempo, timbre, and manner. Possibly the international nature of musical notation was influenced by the fact that medieval neume notation arose at a time when Latin was an international lingua franca, and also by the international nature of late feudal culture, rather than being an internationally-agreed-upon consensus of scientific symbolism. Based on the principles of musical notation, there have been several experimental attempts to construct an international system of dance notation, with signs to designate the position and motions of all parts of the body, with diacritical modifications to indicate tempo and the like. But while the motions of the classical ballet are highly stereotyped, they are semantically meaningless (unlike *natya* dancing in India and Ceylon, and Chinese and other Asiatic theatrics), so that this dance notation is mere *orders to action* like musical notation, with no other semantic content. Western dancing as an art form must appear insipid in its semantic emptiness to an Oriental who is used to articulate literary *meaning* in his dance forms. This is not to deny, however, that Occidental kinaesthetic language *may* be heavily imbued with great subtleties of meaning: the pantomime of the early Charlie Chaplin achieved at least a pan-European understanding and appreciation, while the implicit conventionalizations and stereotypes of Mickey Mouse (a psychiatrically most interesting figure!) are achieving currently an intercontinental recognition and enjoyment.

If all these various ways of *talking* be generously conceded to be purely cultural behavior, surely *walking*—although learned—is a purely physiological phenomenon since it is undeniably a panhuman trait which has brought about far-reaching functional and morphological changes in man as an animal. Perhaps it is, basically. And yet, there would seem to be clear evidence of cultural conditioning here. There is a distinct contrast in the gait of the Shans of Burma versus that of the hill people: the Kachins and the Palaungs keep time to each step by swinging the arms from side to side in front of the body in semicircular movements, but the Shans swing their arms in a straight line and do not bring the arms in front of the body. Experts among the American missionaries can detect the Shan from the Palaung and the Kachin, even though they are dressed in the same kinds of garment, purely from observing their respective gaits, and as surely as the character in a Mark Twain story detected a boy in girl's clothes by throwing a rat-chunker in his lap (the boy closed his legs, whereas a girl would spread her skirt). If an American Indian and an adult American male stride with discernible mechanical differences which may be imputed to the kinds of shoes worn and the varying hardness of the ground in woods or city, the argument will not convince those who know—but would find it hard to describe—that the Singhalese and the Chinese simply and unquestionably just do walk differently, even when both are barefooted. Amazonian tribes show marked sexual contrasts in their styles of walking: men place one foot directly in front of the other, toes straight forward, while women walk in a rather stilted, pigeon-toed fashion, the toes turned inward at an angle of some thirty degrees; it is regarded as a sign of power if the muscles of the thighs are made to come in contact with each other in walking. To pick a more familiar example, it is probable that a great many persons would agree with Sapir's contention that there does exist a peculiarly East European Jewish gait—a kind of kyphotic Ashkenazim shuffle or trudge—which is lost by the very first generation brought up in this country, and which, moreover, may not be observed in the Sephardic Jews of the Iberian Peninsula. Similar evidence comes from a recent news article: "Vienna boasts that it has civilized the Russians . . . has taught them how to walk like Europeans (some Russians from the steppes had a curious gait, left arm and left foot swinging forward at the same time)."[22] The last parenthesis plays havoc with behavioristic notions concerning allegedly quadrupedal engrams behind our "normal" way of walking!

It is very clear that the would-be "natural" and "instinctive" gestures of actors change both culturally and historically. The back-of-the-hand-to-the-forehead and sideways-stagger of the early silent films to express intense emotion is expressed nowadays, for example, by making the already expressionless compulsive sullen mask of the actress one shade still more flat: the former technique of exaggerated pantomime is no doubt related to the limitations of the silent film, the latter to the fact that even a raised eyebrow may travel six feet in the modern close-up. The "deathless acting" of the immortal Bernhardt, witnessed now in ancient movies, is scarcely more dated than the middle-Garbo style, and hardly more artificially stylized than Hepburn's or Crawford's. Indeed, for whatever reason, Bernhardt herself is reported to have fainted upon viewing her own acting in an early movie of *Camille*.[23] There are undoubtedly both fashions and individual styles in acting, just as there are in painting and in music composition and performance, and all are surely far removed from the instinctual gesture. The fact that each contemporary audience can receive the communication of the actor's gestures is a false argument concerning the "naturalness" of that

---

[22] Paula Hoffman, Twilight in the Heldenplatz, *Time*, June 9, 1947, **49:23**, 31. A related kind of motor habit—which is of course conscious—was that of the Plains Indian men who wore the buffalo robe "gathered . . . about the person in a way that emphasized their action or the expression of emotion" (*Handbook of American Indians north of Mexico*, Bulletin 30, Bureau Amer. Ethnol., Washington, D. C., 1907-1910; 2 vols.). For the Amazonians, see Whiffen, *op. cit.*, p. 271.

[23] Maurice Bardèche and Robert Brasillach, *The history of motion pictures* (New York, 1938), p. 130.

gesture: behavior of the order of the "linguistic" (communication in terms of culturally agreed-upon arbitrary symbols) goes far beyond the purely verbal and the spoken.

That this is true can be decisively proved by a glance at Oriental theatrics. Chinese acting is full of stylized gestures which "mean" to the audience that the actor is stepping over the threshold into a house, mounting a horse, or the like; and these conventionalizations are just as stereotyped as the colors of the acting masks which indicate the formalized personalities of the stock characters, villains or heroes or supernaturals. In Tamil movies made in South India, the audience is quickly informed as to who is the villain and who the hero by the fact that the former wears Europeanized clothing, whereas the latter wears the native *dhoti*. But this is elementary: for the intricate *natya* dancing of India, the postural dance dramas of Bali, and the sacred *hula* of Polynesia are all telling articulated stories in detailed gestural language. That one is himself illiterate in this language, while even the child or the ignorant countryman sitting beside one on the ground has an avid and understanding enjoyment of the tableau, leaves no doubt in the mind that this *is* a gestural language and that there *are* sublinguistic kinaesthetic symbolisms of an arbitrary but learnable kind.

Hindu movies are extraordinarily difficult for the Occidental man to follow and to comprehend, not only because he must be fortified with much reading and knowledge to recognize mythological themes and such stereotypes as the *deus-ex-machina* appearance of Hanuman the monkey-god, but also because Americans are characteristically illiterate in the area of gesture language. The kinaesthetic "business" of even accomplished and imaginative stage actors like Sir Laurence Olivier and Ethel Barrymore is limited by the rudimentary comprehension of their audiences. Americans watch enthusiastically the muscular skills of an athlete in *doing* something, but they display a proud muckerism toward the dance as an art form which attempts to *mean* something. There are exceptions to this illiteracy, of course, notably among some psychiatrists and some ethnologists. Dr. H. S. Sullivan, for example, is known to many for his acute understanding of the postural tonuses of his patients. Another psychiatrist, Dr. E. J. Kempf, evidences in the copious illustrations of his "Psychopathology" a highly cultivated sense of the kinaesthetic language of tonuses in painting and sculpture, and can undoubtedly discover a great deal about a patient merely by glancing at him. The linguist, Dr. Stanley Newman, has a preternatural skill in recognizing psychiatric syndromes through the individual styles of tempo, stress, and intonation.[24] The gifted cartoonist, Mr. William Steig, has produced, in *The Lonely Ones,* highly sophisticated and authentic drawings of the postures and tonuses of schizophrenia, depression, mania, paranoia, hysteria, and in fact the whole gamut of psychiatric syndromes. Among anthropologists, Dr. W. H. Sheldon is peculiarly sensitive and alert to the emotional and temperamental significance of constitutional tonuses.[25] I believe that it is by no means entirely an illusion that an experienced teacher can come into a classroom of new students and predict with some accuracy the probable quality of individual scholastic accomplishment—even as judged by other professors—by distinguishing the unreachable, unteachable *Apperceptions masse*-less sprawl of one student, from the edge-of-the-seat starved avidity and intentness of another. Likewise, an experienced lecturer can become acutely aware of the body language of his listeners and respond to it appropriately until the room fairly dances with communication and countercommunication, head-noddings, and the tenseness of listeners soon to be prodded into public speech.

[24] Stanley S. Newman, Personal symbolism in language patterns, *Psychiatry,* 1939, **2,** 177-184; Cultural and psychological features in English intonation, *Trans. N. Y. Acad. Sci.,* 1944, ser. II, **7,** 45-54; (with Vera G. Mather), Analysis of spoken language of patients with affective disorders, *Amer. J. Psychiat.,* 1938, **94,** 913-942; Further experiments in phonetic symbolism, *Amer. J. Psychol.,* 1933, **45,** 53-57; Behavior patterns in linguistic structure, a case history, in *Language, culture and personality, Essays in honor of Edward Sapir* (Menasha, Wis., 1931), pp. 94-106. The Witoto and Bororo have a curious motor habit: "When an Indian talks he sits down—no conversation is ever carried on when the speakers are standing unless it is a serious difference of opinion under discussion; nor, when he speaks, does the Indian look at the person addressed, any more than the latter watches the speaker. Both look at some outside objects. This is the attitude also of the Indian when addressing more than one listener, so that he appears to be talking to some one not visibly present." A story-teller turns his back on the listener and talks to the wall of the hut (Whiffen, *op. cit.,* p. 254).

[25] W. H. Sheldon, *The varieties of temperament* (New York & London, 1942). The argument of one variety of athletosome or somatotonic scientist that Sheldon is unable or unconcerned to muscle his findings into manageable, manipulable statistical forms wherewith to bludgeon and compel the belief of the unperceiving, is of course peculiarly irrelevant. The psychiatrist soaked in clinical experience is similarly helpless in his didactic relations with a public which either has not, or cannot, or will not see what he has repeatedly observed clinically.

The "body language" of speakers in face-to-face conversation may often be seen to subserve the purposes of outright linguistic communication. The peoples of Mediterranean origin have developed this to a high degree. In Argentina,[26] for example, the gesture language of the hands is called "ademanes" or "with the hands." Often the signs are in no need of language accompaniment: "What a crowd!" is stated by forming the fingers into a tight cluster and shaking them before you at eye level; "Do you take me for a sucker?" is asked by touching just beneath the eye with a finger, accompanying this with appropriate facial expressions of jeering or reproach as the case might be; and "I haven't the faintest idea" is indicated by stroking beneath the chin with the back of the palm. One Argentine gentleman, reflecting the common notion that *ademanes* have the same vulgarity and undignified nature as slang—appropriate only for youngsters or lower-class folk—nevertheless, within five minutes of this statement, had himself twirled an imaginary moustache ("How swell!") and stroked one hand over the other, nodding his head wisely ("Ah ha! there's hanky-panky going on there somewhere!"). Argentine gesture-language is nearly as automatic and unconscious as spoken language itself, for when one attempts to collect a "vocabulary" of *ademanes,* the Argentine has to stop and think of situations first which recall the *ademanes* that "naturally" follow. The naturalness of at least one of these might be disputed by Americans, for the American hand-gesture meaning "go away" (palm out and vertical), elbow somewhat bent, arm extended vigorously as the palm is bent to a face-downward horizontal position, somewhat as a baseball is thrown and in a manner which could be rationalized as a threatened or symbolic blow or projectile-hurling) is the same which in Buenos Aires would serve to summon half the waiters in a restaurant, since it means exactly the opposite, "Come here!" When the Argentines use the word "mañana" in the familiar sense of the distant and improbable future, they accompany the word by moving the hand forward, palm down, and extending the fingers lackadaisically—a motion which is kinaesthetically and semantically related perhaps to the Argentine "come here!" since this symbolically *brings,* while "mañana" *pushes off.* Kissing the

[26] Arthur Daniels, Hand-made repartee, New York *Times,* October 5, 1941.

bunched fingertips, raising them from the mouth and turning the head with rolled or closed eyes, means "Wonderful! Magnificent!," basically perhaps as a comment or allusion to a lady, but in many remotely derived senses as well. "Wonderful!" may also be expressed by shaking one of the hands smartly so that the fingers make an audible clacking sound, similar to the snapping of the fingers, but much louder. But this gesture may signify pain as well as enjoyment, for if one steps on an Argentine's toes, he may shake his fingers as well as saying "Ai yai!" for "ouch!" The same gesture, furthermore, can be one of impatience, "Get a move on!" Were one to define this gesture semantically, then, in a lexicon of *ademanes,* it would have to be classified as a nondescript intensificative adverb whose predication is indicated by the context. In fast repartee an Argentine, even though he may not be able to get a word in edgewise, can make caustic and devastating critiques of the speaker and his opinions, solely through the subtle, timed use of *ademanes.*

A study of conventional gesture languages (including even those obscene ones of the *mano cornuta,* the thumbed nose, the *mano fica,* the thumbnail snapped out from the point of the canine tooth, and so forth,[27] as well as those more articulated ones of the Oriental dance dramas), a study of the body language of constitutional types (the uncorticated, spinal-reflex spontaneity and *legato* feline quality of the musclebound athletosome, his body knit into rubbery bouncing tonuses even in repose; the collapsed colloid quality of the epicurean viscerotonic whose tensest tonus is at best no more than that of the chorion holding the yolk advantageously centered in the albumen of an egg, or the muscle habituated into a tendon supporting a flitch

[27] The only place I have seen this discussed recently is in an article by Sandor Feldman, The blessing of the Kohenites, *American Imago,* 1941, **2**, 315-318. In the same periodical is an exquisitely sensitive interpretation of one person's interpretation of the signs of the zodiac in terms of positions and tonuses of the human body (Doris Webster, The origin of the signs of the zodiac, *ibid.,* 1940, **1**, 31-47). Other papers, of the few which could be cited with relevance to the present problem, would include: Macdonald Critchley, *The language of gesture* (New York, 1939); G. W. Allport and P. E. Vernon, *Studies in expressive movement* (New York, 1933); F. C. Hayes, Should we have a dictionary of gestures? *Southern Folk-Lore Quarterly,* 1940, **4**, 239-245; Felix Deutsch, Analysis of postural behavior, *Psychoanal. Quart.,* 1947, **16**, 195-213; Paul Schilder, *The image and appearance of the human body,* Psyche Monographs (London, 1935); Th. Pear, Suggested parallels between speaking and clothing, *Acta Psychol.,* Hague, 1935, **1**, 191-201; J. C. Flugel, On the mental attitude to present-day clothing, *Brit. J. med. Psychol.,* 1929, **9**, 97; La Meri, *Gesture language of the Hindu dance* (New York, c. 1940); Rudolf von Laban, *Laban's dance notations* (New York, c. 1928).

of bacon; and the multiple-vectored, tangled-stringiness of the complex "high-strung" cerebrotonic, whose conceptual alternatives and nuances of control are so intricately involved in his cortex as to inhibit action), and the study of psychiatric types (the Egyptian-statue grandeur and hauteur of the paranoiac's pose; the catatonic who offers his motor control to the outsider because he has withdrawn his own executive ego into an inner, autistic cerebral world and has left no one at the switchboard; the impermanent, varying, puppet-on-a-string, spastic tonuses of the compulsive neurotic which picture myotonically his ambivalence, his rigidities, and his perfectionism; the broken-lute despair of the depressive; and the distractable, *staccato,* canine, benzedrine-muscledness of the manic)—all might offer us new insights into psychology, psychiatry, ethnology, and linguistics alike.

NAVAHO MOTOR HABITS

By FLORA L. BAILEY

NAVAHO MOTOR HABITS

By FLORA L. BAILEY

THE importance of motor habits as one aspect of the social heredity of particular groups has long been recognized and in various publications[1] may be found valuable information. As yet, however, nothing approaching a full systematic description of the motor habits of a single people has been presented.

The material to be considered here was obtained either by direct observation in the field or by the use of a motion picture camera which allowed closer analysis later. Observations were made over a period of three summers within a radius of ninety miles from Gallup, New Mexico. An actual count of cases was made for each habit until it became obvious that for certain movements a generalization could be formed.[2]

A detailed description of each motor habit will be made. Only those will be mentioned which differ from the characteristic method of performing a similar movement in white (American) culture, or are movements which are not paralleled at all. Since the investigator has had very little field contact with other Indian cultures, but is accustomed to observing and analyzing active physical movement in our culture, she has chosen this as a basis of selection rather than making a comparison with other primitive groups.

General Observations: One of the most striking differences first noted between Navaho movement and that of the white American is the smoothness and flowing quality of the action. Briefly, the Navaho gestures and moves with sustained, circular motions rather than with the angular, staccato movements characteristic of white culture. This is especially noticeable in the gestures accompanying a narration, as will be mentioned later. However, it is present to a certain degree in all movement. Perhaps it is an evidence of greater relaxation and less strain, resulting from a more leisurely mode of living than American culture affords. This explanation seems inadequate taken alone and there are doubtless other contributing factors, some of which will be mentioned as they occur in context.

The motor habits which have been listed on the following pages are grouped arbitrarily under such headings as Personal Habits, Social Habits, Work Habits, and the like. This has been done for ease of reference.

PERSONAL HABITS

Eating: Eating is done daintily and slowly. Small pieces of meat are

---

[1] See, for example: Franz Boas, *Primitive Art* (Oslo, 1927), page 144–149; and Leslie Spier, *Havasupi Ethnography* (Anthropological Papers of the Museum of Natural History, Vol. XXIX, Part I), page 136.

[2] I took the observation of a *minimum* of four cases as the basis for a generalization.

cut from the larger one with a knife. Sometimes the meat is grasped with the teeth and the left hand and cut off close to the mouth with a knife, cutting toward the face. Sometimes the piece desired is held in the fingers and severed from the main portion with the knife moving away from the body.

Bread is torn into small pieces and dipped in broth or stew, then placed in the mouth with thumb on top and fingers underneath pointing toward the face. If only bread and coffee is served, the bread is torn into small pieces and eaten with the fingers.

Coffee is stirred with a spoon to dissolve the sugar. The spoon may be left in the cup while drinking or laid on the ground or the low table.

Spoons are used with both the common dish of stew and with individual dishes. They are laid to rest on the edge of the bowl between bites, the handle on the table and the bowl on the lip of the dish facing down. Sips are taken either from the side or the end of the spoon.

After eating, the fingers are wiped on a handerkchief, a towel, a rag, or the clothing.

Melons are eaten by biting from the slice. No one has been observed eating a melon with an instrument, although women have been seen using the forefinger of the right hand, knuckles up, with a soft, pushing motion away from the body to dig the melon pulp onto the fingernail. One three year old child ate this way also.

In drinking hot coffee the breath is drawn in with the liquid making a loud sucking noise. This cools the drink which is served scalding hot. At the close of the swallow the breath is expelled in an aspirated "aaahh" which indicates appreciation and enjoyment. Older men almost invariably exaggerate this action.

After finishing the coffee the dregs are tossed into the fire. One informant explained this by saying that "you must always feed the fire for it is hungry," adding "when there is grease left over, it is dumped into the fire too."

Sleeping: The usual sleeping posture for men and women is on one side, legs either flexed or extended, the head well supported. Occasionally during the daytime a man sleeps on his back, ankles crossed, hands beneath his head, hat over his eyes.[3] Children sleep with little support for the head. If they are put to bed by an adult, they are usually placed on their sides with knees slightly flexed. They often turn to sleep on their backs, arms outflung. No child has been observed sleeping with a hand under the face or head as white children often do. A young child is often put to bed in a

---

[3] Gladys A. Reichard, *Dezbah* (Augustine, New York, 1939), photograph opposite page 46.

nursing position. The mother lies on one side facing the child and cradles him in her arms, patting his buttocks as he nurses. She remains beside him until he falls asleep. One man was observed putting a small boy to sleep by assuming the same nursing position a woman would. The child immediately fell asleep and then the man left him.

Combing Hair:[4] A grass brush is used for the hair. It is grasped in the fist with little finger down and thumb up. The stubby end is held pointed toward the ground and the longer fan-shaped end up. The stubby end is used for the brush and the strokes are from the forehead back. When the hair is brushed smoothly back from the face and ears, the hair cord is tied twice around the hair making a "tail" which starts high on the occipital bone. The ends of the cord are then brought around and held crossed between the teeth. The "tail" is then held straight up in the air and turned down over and over toward the head. The roll thus made is about four to six inches long. When the roll reaches the head, the cord is removed from the mouth and wound around the middle of the roll leaving the ends to hang down when tied. When the roll is spread out at top and bottom, it assumes the butterfly shape commonly seen on both men and women.

The hair is usually combed over a towel or other cloth in which the combings are collected. They are later burned.[5] Frequently one person combs the hair of a second, and in exchange has hers combed. Sometimes while brushing the hair in this manner, each person will search the other's head carefully for lice.[6]

Spitting: Spitting is a constant practice of both men and women and is not presumably done for relief as much as from habit. The men spit through the teeth in the general direction of the fire, if indoors. The women lean over and spit onto the ground, using their fingers to wipe the mouth free of saliva. A little dirt is then brushed over the mucus with the fingers to cover it up. Singers lean forward, making ready to spit, and at a suitable moment in the song they spit and continue singing. Though they may miss a word, the song continues on a humming note and there is no break in the rhythm. One singer was timed as spitting about every two minutes. Distances vary from two to six feet.

Ablutions: The hands are washed frequently, using a wringing motion, and are dried on a towel or the skirt. The face is washed by splashing

---

[4] *Ibid.*, frontispiece

[5] Kluckhohn in his field notes mentions this disposal as a protection against witchcraft.

[6] W. W. Hill states in his field notes that lice are either picked from the hair with the fingers or the individual will wash his hair and then enter the sweat house.

water on it with the hands. The body is washed in the same way, then rubbed down and dried either on a towel or the skirt.

Nose-Blowing: The nose is generally cleared between the thumb and forefinger and the mucus either shaken off or wiped on the clothing. Sometimes a handkerchief is used.

Urinating:[7] Young men urinate standing. Older men kneel either on one knee or both knees to urinate. Young women squat down, and older women urinate standing.

Pointing: One informant states that you point with your lips if you are close enough so a person can see you, but if he is too far away to see your lips, you point with your finger and hand. You may also point with your head. It will be noted that many individuals have been observed pointing with both the lips and finger simultaneously, although the most common method is with the lips alone. These are pursed forward, slightly open, in the direction to be indicated. The head is slightly turned in this direction also. In the word *ńlêidi*, meaning "over yonder," the pursing is done in the middle of the word, making the word and gesture coincide. When pointing with the forefinger, the whole arm is extended and the motion seems to originate at the shoulder and flow out.

The head is nodded to indicate "yes" and shaken from side to side to mean "no."

The youngest age at which a child was observed pointing with his lips was three years. He also pointed with his finger at the same time.

Shy-Gesture:[8] The women make a characteristic gesture of shyness in which the mouth is covered by the hand, especially when smiling or laughing. The right elbow is held close to the side, slightly in front of the body, the right finger tips rest on the cheek under the right eye, and the outer edge of the little finger covers the upper lip with the thumb along the chin line on the right side of the face.

Variations of this gesture are often seen. The left hand is sometimes used; sometimes when sitting the elbow is rested on the upraised knee and the hand rests across the mouth in repose; sometimes the position of the hand is with the forefinger up and the thumb pointing toward the eye on the same side as the hand used; and sometimes the elbow is raised instead of being held close to the side. In any event, the gesture of the hand over mouth is uniformly characteristic of shyness in a women. It has been observed, however, that among a group of women who are close acquaintances

---

[7] Information offered by W. W. Hill from his field notes.
[8] Reichard, 1939, *op. cit.*, photograph opposite page 39.

or relatives, when no outsider is present the shy-gesture is entirely omitted and laughing is done openly and without embarrassment.

Cigarette Smoking: The most frequently observed method of holding a lighted cigarette is with the thumb underneath and the fingers on top. Men hold the cigarette to their mouths in this position with one hand and draw a light from either a match or a glowing coal held in the other hand. Often the cigarette is held away from the mouth while the match is applied. Old women in particular often hold the cigarette in this fashion and light it by alternately holding the cigarette in contact with the flame, then putting it to the mouth and puffing on it. Younger women usually light the cigarettes in the same way as the men. They do not inhale, however, but puff with short, rapid breaths in an awkward manner. One very old woman was observed inhaling deeply, then blowing the smoke in a long stream out the left corner of her mouth. Occasionally a person is seen holding a short stub to the mouth between the thumb and forefinger, fingers up and pointing toward the lips, the lighted end shielded by the cupped hand.

Cigarettes are usually thrown away without being extinguished, or tossed into the fire if convenient. One woman was observed extinguishing a half-smoked cigarette by brushing it gently on the ground. She then put it away to finish later, remarking that it made her dizzy to smoke.

An old woman was noticed holding a cigarette between the inner sides of the first and second fingers in the customary white fashion, but with the palm away from the body instead of facing it, the lighted end of the cigarette being on the palm side of the hand.

Sitting Postures[9]—Children: Boys often drop down on one knee, toe turned in, and sit on the inner border of the foot. The other foot rests flat on the ground, knee up, with the boy's arm resting on the knee, hand dangling. They also imitate the men's sitting postures or squat as described for small girls.

Small girls frequently squat with both feet flat on the ground, knees up, heels almost touching each other. Older girls imitate women's positions.

Once a twelve year old girl squatted on her heels and rested her buttocks against a wall for support.

Women:[10] Often they sit[11] on one hip, both legs turned back toward

---

[9] W. W. Hill, *An Outline of Navaho History, Ethnography, and Acculturation* (Federal Writer's Project, mss. 1937). In referring to the sitting postures of Navaho men and women Hill indicates that they have a precedent in the positions assumed by two of the legendary deities, thus making them traditional.

[10] W. W. Hill in his field notes says that people would talk about a woman who sat in the sitting posture of a man and wonder what was wrong with her. They would say "I wonder what she is doing that for? That doesn't look like a woman."

[11] Reichard, 1939, *op. cit.*, photographs opposite pages 102, 119, and 63.

the opposite side and feet tucked under the body. Frequently the upper knee is slightly lifted and crossed or drawn forward to rest somewhat over the lower knee. The weight of the body is supported on the heel of the hand with the arm extended down at the side, or the woman slumps forward with her arms around her knees.

Sometimes the feet are stretched out straight in front, weight on the buttocks, the body held erect with slightly rounded shoulders and slumped torso. When the hands are placed, palm up, on the thighs, it completes the ceremonial sitting position for women. This is occasionally varied by crossing the ankles.

Another position is to sit on the buttocks, one leg rotated outward and knee flexed to bring the foot close to the body, the outer border of thigh and leg resting on the ground. The other foot is placed flat on the ground with the knee raised. The forearm on this side of the body rests on the upraised knee with the hand dangling.

Women sometimes sit on the inner border of one foot as described for men, with the other knee flexed, or with leg extended straight forward.

Men: They squat on their heels, back and buttocks resting against a wall or other support about twelve inches from the ground. The arms rest on the knees.

Sitting cross-legged, "tailor fashion," hands placed, palms up, on thighs, is the ceremonial posture for the men.[12]

A half-lying position on the back is often taken with shoulders and head propped up against a wall or other support, knees bent up and out, feet flat on the ground, and hat resting on one knee, or over the eyes (if sleeping). Great variations are seen in the reclining angle in this position.

They squat "cowboy fashion" with one foot flat on floor, knee raised while the other foot with heel raised allows the weight to rest on the toe. The buttocks rest on the raised heel of this foot. This is a common eating position for men.

Both men and women often sit on the inner border of one foot, the leg being rotated out and flexed under the body, while the other foot rests knee upraised and foot flat on ground.

Walking:[13] Both men and women walk erect with a long stride in a relaxed, loose-limbed fashion, feet toeing ahead, legs swinging freely from the hips. There is a difference in movement varying with the type of shoes worn. The moccasin-clad individual walks with more grace and ease than one clad in store shoes. This is understandable and probably due to the

---

[12] Reichard, 1939, *op. cit.*, photograph opposite page 142.

[13] W. W. Hill says in his field notes that informants maintain you must walk side by side with your guest, preferably with arms about each other's shoulders.

discomfort of badly fitting shoes. Cowboy boots on the men produce the peculiar walk noticed among white men wearing similar footgear; a slight flexion of the knees, tilt of the pelvis, a shortened, rather awkward gait due to high heels.

The shoulders of both men and women tend to droop slightly forward, arms hanging loosely. The head, however, is erect and the chest does not take the same hollow concavity that drooping shoulders and fatigue posture produces in whites. The effect is that of ease, relaxation, and control in the walk. Especially in the women, aided no doubt by the graceful swing of a long, full skirt, it is most pleasing. It contrasts decidedly with the gait of the Pueblo woman which, viewed from the rear, is a waddle from side to side.

Horseback Riding:[14] Both men and women ride with long stirrups and stand in them when trotting. The reins are held in either hand, or both, in no set style. The elbows lift away from the sides and move with the motion of the horse. The position is easy and relaxed, and the body slumps slightly. In riding bareback or when racing, the knees are flexed and the horse gripped with them. Then the body is bent forward, reins in the left hand, while the whip is handled with the right. The quirt is brought down across the horse's flanks with a rhythmic lift and swing as the animal runs. Both men and women show expert horsemanship which appears efficient and utilitarian rather than conforming to a set style.

SOCIAL HABITS

Handshake:[15] A Navaho handshake is a limp clasp and release of the right hands. There is no motion up and down, no grip. In greeting a family, precedence is given to the older members present. You then shake hands as desired with others assembled. On one occasion three young men were observed to enter a cook-shade, squat and eat, then having finished turn to greet those near by with handshakes although previously ignoring them.

Stylized Embrace:[16] The right hands are grasped as for the handshake, but the left arm is lifted and placed around the neck of the person greeted, to lie loosely across the shoulders. The head is inclined forward toward the right shoulder of the person who is so embraced. This has been observed

---

[14] Reichard, 1939, *op. cit.*, photographs opposite pages 94 and 110.

[15] Hill states in his field notes that handshaking is a custom taken over from the Mexicans. It was customary in the old days to embrace.

[16] Hill describes the embrace with one hand over and on the back of the shoulder, the other under the opposite arm. The older people of a tribe or a father and son will embrace each other on meeting.

on only one occasion, at which time four older members of a family thus greeted a long absent friend.

Sitting Position in the Hogan: Any of the positions and variations described in the section on sitting postures may be assumed. It is common to see women sitting on the north of the door, men at the south. On ceremonial occasions this location is prescribed. The head of the house frequently sits to the west of the fire. Etiquette demands that no one remain standing for any length of time in the hogan.

Measuring[17] *and Counting:* One woman was observed indicating the number of years by the following gesture: the four fingers of the right hand were held before the chin, about ten inches away, close together and straight with the fingers pointing up, palm toward the face. Her thumb was curled across the palm, hidden, and the fingers bent at phalangealmetacarpal joint. She emphasized the years by a slight motion as if pushing toward the listener with the back of her hand several times.

The amount of weaving finished in one day (one inch) was indicated by placing the first two fingers of the right hand on the ground perpendicular to it and facing the body.

One old man indicated the number ten by using the fingers of both hands extended with the palms out, thumbs together, and backs toward his face. Later he was observed showing other numbers in the same manner by extending the required number of fingers.

A woman counted three by holding up the last three fingers of her right hand in front of her face, palm toward the face.

Gesture of Appreciation: Men in relating a story, or in showing appreciation of one, bring the hands together once, slowly and silently, in a clapping motion and softly rub them against each other a few times. The motion is preceded by a fairly wide swing and has the effect of an exclamation such as "that's a good one!"

Story Telling: In the process of relating a story certain things are noticed. The movements used for illustration are graphic and almost continuous, having a slow, liquid character. The motions are smooth, relaxed,

---

[17] Hill in his field notes lists the following measurements which are indicated by motions: one, two, three, four, and five finger widths (five finger widths being measured from the knuckle of the thumb over the hand): the spread from thumb to index finger, middle finger, ring finger, or little finger: the distance between the second and third joints of the little finger (this measurement is used in cutting prayersticks as are also three and four finger widths): from the tip of the little finger to the elbow: from the middle of the chest to the tip of the little finger, arm outstretched: from the nipple of the opposite breast to the tip of the middle finger of the outstretched arm (this measurement is used in marking the length of a bow): and the span between the arms.

yet sustained, and only in a few spots are they staccato. They are in general circular not angular movements and there is much swaying of the hands back and forth. Movements seem to originate at the shoulder and extend through the fingers.

To indicate direction or distances traveled, the forefinger is sometimes pointed toward the ground away from the body with the other fingers loosely folded under. In this sort of motion an "attack" is made in the air at a point near the body, then a sustained movement is made forward as if drawing a line in space, and it is finished with a final "attack" at arm's distance to indicate destination reached. A continuous motion away from the body, finally drifting away in space, is used to indicate extreme distance or time.

One man held the left forefinger pointed away from his body parallel to the floor with the thumb side up. The right forefinger was then used to point along this surface with an emphasizing gesture. A point at the base of the left finger was first indicated, then the hand moved along to draw a line toward the tip, indicating branches to right or left. In this motion the right finger was held fairly flat, palm down, and the tip shoved along in advance of the hand.

One man was seen describing a prairie dog in the following manner: The curved index finger of the right hand was held erect, other fingers clenched in a fist, with the thumb side toward the body. The onlooker saw the lifted finger resemble a prairie dog sitting up by his hole.

Dancing:[18] In the "Girls' Dance" portion of the Enemy Way, dancing in couples occurs. The usual method of approach is for the girl to clutch the man securely and force him to follow her into the circle of dancers around the fire. (One informant states that school-boys sometimes ask a girl to dance using the phrase "May I dance with you." This would be entirely outside tradition.)

The motion of the couple dance is a shuffle of the feet. Partners stand side by side, facing the same direction, with inside arms linked or placed around each other's waists. They move slowly forward around the circle. Sometimes couples stand face to face with hands on each other's waists, then turn in place with shuffling steps. Occasionally a couple embellishes the steps by moving forward with a slow, accented skip. In one instance a polka was observed, at Smith Lake.

Bathing Infant: The woman sits on the ground, her legs folded under

---

[18] B. and M. G. Evans in their book *American Indian Dance Steps* (New York, 1931) have described in detail certain Navaho dance steps. Therefore, only the "Girls' Dance" will be described here.

to the right. A wash basin is placed on the ground in front of her knees and the water tested with the right thumb to ascertain correct temperature. The infant is held face up on her lap, its head supported by her left hand placed under its back and neck. The head is pointed toward the woman's knees, buttocks resting high on right thigh, legs hanging over right thigh toward the woman's hip. Using the right hand she first bathes the head. One informant stated that all Navaho women washed the infant's head first, but she didn't know why. Either a cloth or the bare hand is used. The rest of the body is bathed very sketchily compared with the thoroughness of the shampoo. The infant is sometimes supported upright in the basin after the shampoo and his bath completed in this position.

Nursing Habits: A variety of nursing positions have been observed, differing with the age of the child and the inclination of the mother. A common practice when the child is in the cradle board is to lay the board across the lap and let the child turn its head to nurse.

One two year old boy (naked) left his play, approached his mother from the left side, lay face down across her left thigh with legs sprawled on the floor, and reached for her breast. She gave him no help other than uncovering the breast and laying her left hand across his back.

A fourteen month old child was nursed while lying in her mother's lap, supported face up by her mother's arm, head resting on the curved elbow. In this same position another mother nursed her five month old baby. She alternated breasts and supported the nipple in position by holding it between the first two fingers of the right hand, palm flat against the breast.

An eighteen month old boy kneaded his mother's breasts vigorously each time he nursed, though she stated that he did not bite. Other nursing positions of the same boy are as follows: standing in front of the kneeling mother; standing in front of mother who is leaning over a bench working; lying on his face across mother's right knee from the right side to reach the corresponding breast, and vice versa; lying on the floor, or mother's lap, face up and supported by her arm.

Nursing positions in order to put a child to sleep have already been mentioned.

Jouncing Infant: Children are frequently fondled and jounced by members of the family either to prevent crying or simply to show affection. One woman held her five month old daughter on her lap and bounced her up and down at length, kissing her regularly and in rhythm as she moved.

An old woman was seen jouncing her grandchild in this position: legs folded under her to the right, the baby seated on the right thigh with its back toward the woman's body and supported by her hands under its arms. Moving her right thigh in a series of alternate abductions and adductions she produced a rhythmic movement which bounced the baby. When she stopped the movement she placed a sheepskin under the right knee to raise the leg, and held the baby in the same position at rest.

A girl of ten was seen rocking her sister in a cradle board by using the wagon seat as a lever for a see-saw and balancing the cradle over its edge while teetering it back and forth.

Another woman held her infant supported under its arms in a standing position in her lap facing away from her body. She bounced the child up and down, hissing slightly to quiet her.

A rocking motion was produced for one baby in a cradle board in this manner: the cradle was supported upright on the ground at the women's left by her left hand. Her legs were folded to the right, and the cradle lay back against her left shoulder, baby's face away from her body. In this position the cradle was slowly swung from side to side using the base as a fulcrum. The left arm swung parallel to the ground. The woman crooned "ñcah, ñcah" (crying, crying) while rocking the infant.

Another woman while fondling and jouncing her child in her arms frequently kissed it on the mouth. This is the only occasion when any Navaho has been seen kissing another directly.

Wrapping and Tying Cradle Board: If the woman is right handed, the baby's head is placed toward her left when he is laid on the board for wrapping. If she is left handed, the head is placed on the right side. In either case, the wrapping is begun from the infant's right to left (i.e., the right handed woman wraps away from her body in the first motion, the left handed one toward her in the first motion).

The cradle is prepared by first placing oblong pieces of torn flour sacking over the board in several layers. Triangularly folded pieces are next laid down with the apex toward the foot of the board. A U-shaped pad is fastened at the upper end of the board to support the head and neck, the opening of the U at the top. Within this space is laid a small, flat pillow. Powder is placed on the triangular diapers and the baby laid on top. The diapers are folded with motions similar to those used by white women in the same situation, although they are not pinned in place. The apex of the triangle is brought up between the legs and the two side pieces folded across the abdomen to meet it. The rectangular pieces are then folded over the child's body. The end on the child's right is folded over the right arm, which is extended close to the side, and tucked under the left arm. The left end is then pulled across the left arm, which is extended at the side, and **brought across the body to the right, then continued entirely around the** body. This confines both arms. The lower ends are then tucked under the feet and the buckskin lacing on the board is tied. The lacing is always begun at the baby's right shoulder and crosses his body to the left on the first motion. When the lacing reaches the foot, the foot-board is raised in place and fastened. This completes the procedure.[19]

Diapers are only used when the baby is in the cradle. Some women use shredded cliff rose bark (ʔawé·čál) for diapers. They wrap some of the bark in a small blanket and use it as a pad under the infant. This bark can be dried in the sun and used over again. They report that no odor exists. Many women, however, use flour sacks or other cloth rather than the older bark diapers. Women who use the bark have two bags full, alternating them. One dries while the other is in use.

Methods of Entry into the Sudatory and Sitting Postures:[20] Only three women have been observed in a sweat house. They entered by sitting down at the door and hitching themselves inside feet first, assisted by the hands. The cramped space may have been responsible for this method of entry, as well as the fact that the floor was about six inches lower than the doorsill. The women sat on the south side of the sudatory, legs folded under them, two to the right and one to the left.

Exit from the house was made head first, crawling out with the weight supported on the hands and knees.

WORK HABITS

Shearing Sheep: Women do the shearing with occasional help from the young men. Shears are held in the right hand, the thumb along one side of the shears, the fingers closed across the opposite blade. The shears may be held with the fingers up, the knuckles up, or the thumb up according to the individual's convenience. The wool is pulled away with the left hand as the cutting progresses. Positions the shearer may take are sitting with the right knee up; with legs folded under to one side; on both heels; kneeling; or standing.

After the sheep is caught and thrown on one side, the feet are tied together, crossed at the first joint with the upper rear foot on top, then the upper front foot, the lower rear foot, and lastly the lower front foot. The rope which ties the feet together is wound from front to back around the ankles, (i.e., from top to bottom toward the animal's body).

Shearing begins from the rear, either at tail or legs, and continues to-

---

[19] Reichard, 1939, *op. cit.*, photograph opposite page 70.

[20] Wyman in a personal communication states that men's sitting postures are not unusual, simply being adapted to the small space (i.e. with knees drawn up).

ward the head. As one side is completed, the shearer stands behind the sheep and throws it on its other side by reaching across the body and pulling the legs over. When both sides are finished, the legs are untied and the stomach clipped. The sheep is held between the shearer's own legs. The shearer places her left leg across the sheep's neck and upper front leg, then holds the upper rear leg firmly in her left hand and manipulates the shears easily with the right hand. When all wool is clipped, the sheep is released. Informants state that it is hard work to shear since the second knuckle of the thumb and the palms get sore.

Carding Wool: The following description of carding is of the process as done by an older woman who was an expert. It is typical of the motions made by any woman who cards. The informant was seated with legs folded under her to the right, holding one carding comb in the left hand with the thumb up and fingers on top. The tip of the handle was away from the body while the comb was toward the body and rested against the left knee, slanting slightly toward the floor so the dirt could fall into her lap. Her right hand held the second comb with knuckles up and first finger on top to brace it. The prongs were turned down facing away from the body. In this position the prongs of the two combs pulled across each other.

Mixed sheep and mohair wool (for warp) was placed on the upper comb (the one held in the right hand) and arranged carefully. The comb was lying in a rest position in her lap face up at this time. The upper comb was then turned from this rest position and pulled down with easy, short movements against the teeth of the lower comb. A stroke about six inches in length was used. After four or five strokes the lower comb was turned by means of an inward rotation of the left forearm so that the handle pointed toward the body with the comb teeth down. At the same time the upper comb was turned over by an outward rotation of the right forearm away from the body with the teeth up. This reversed their relative positions. On every turn of the combs, that is after each four or five strokes, the direction of pull of the upper comb was changed, alternating from a pull toward the body to one away from it. This made the finished wool come out in long, smooth rolls. On the final stroke it was reversed twice in succession before removing it from the comb. The wool turns up on alternate combs after each reversal and brushing.

The informant stated that most women in carding followed these same motions, although two women whom she knows are left-handed and card in directly the opposite manner. She demonstrated a reversal of the process.

Women were observed carding in various sitting positions, right legs under; left legs under; kneeling; and feet straight ahead. One woman used the first two fingers of the right hand on top of the comb to brace it instead of one.

Informants state that carding makes blisters on the palms of the hands and on the right forefinger.

Spinning:[21] Sitting positions for spinning vary in the usual manner, feet fully extended; legs folded under left; legs folded under right; feet extended, ankles crossed, etc. With the spindle held in the right hand, the wool is fastened by sticking the tip through the end of the wool and giving the spindle a twirl. In spinning, the short end of the shaft rests on the ground at the right side, eight to ten inches from the thigh, and the longer end is held between the right thumb and fingers.[22] It is twirled by pulling the fingers of the right hand across the thumb. The right arm is held out at the side, elbow high and bent forward and slightly down with wrist curved down and fingers pointing toward the ground. As the spindle fills with yarn it becomes heavy, and the position and motion is somewhat changed. It becomes a pulling toward the body with the palm of the hand, from the palm toward the fingers. Then catching the spindle in the V formed by the thumb and first finger, it is pushed back to the original forward position for the next stroke. The motion appears to be a pulling and pushing on the right thigh which keeps the spindle rolling.

The left hand is held about eighteen inches from the spindle tip, fingers up, yarn grasped between first two fingers. The spindle is twisted two or three times with the right hand, then held firm in the right fist while the yarn is pulled in quick, elastic jerks away from the tip six to ten times. If the yarn needs special attention at some point, it may be pulled and straightened there with the fingers, laying the spindle down or holding it in the palm of the right hand.

At the end of each group of pulls the left arm is nearly straight out at the left side, shoulder high. The right hand holding the spindle is stretched out at the right side. The base of the spindle where it rests on the ground is used as a fulcrum. At the end of the stretch, the yarn is relaxed slightly, the spindle turned out away from the body and given a swift reverse twirl which winds the yarn to the base of the spindle against the whorl. The process is then repeated until the spindle is full of yarn when it is removed and wound.

Warp yarn must be spun twice to make it finer and firmer.

Weft yarn is spun either from carded wool bats overlapped at the ends

---

[21] Gladys A. Reichard, in her book *Navaho Shepherd and Weaver* (New York, 1936) describes in detail the many processes involved in weaving, and gives an accurate analysis of certain specific movements. [22] Reichard, 1939, *op. cit.*, photograph opposite page 71.

as the spinning begins, or from uncarded wool which is especially fine and is stripped off in long lengths about two inches wide.

One woman was observed spinning with a longer than average spindle whose whorl was six to eight inches from the lower end. She rolled it against the right thigh in the manner described above, but laid the spindle down resting against her leg and used both hands to manipulate the yarn.

Another woman spun with great dexterity, using the spindle in the usual manner but frequently turning her left hand thumb down to pull the wool against the heel of her thumb instead of always against her fingers. At the same time she grasped the spindle firmly with the right fist, little finger down, to hold the yarn firm on the strong pull. She shortened the yarn with her left hand by winding it from the thumb to the little finger and back several times, "spider-ing down" toward the spindle. Then she let it out again and continued the stretching process. She continually carried quite a bit of wool in her left hand.

Winding Wool: Two methods of winding yarn have been observed. One method is to sit with feet extended forward, hold the yarn in the left hand and wind it, with the right hand, around the left hand and the extended left foot, working clockwise.

One woman while winding in this manner was assisted by her husband who held the tip of the spindle, the base resting in a basin, and allowed the yarn to be drawn from the spindle by the woman. The spindle twirled neatly in the basin.

Another woman drew the yarn from the spindle by curving her left arm across her body, resting the spindle in the curve of the elbow, and holding the wool in her left hand. With her right hand she wound the wool off the spindle into a ball. The winding motion was away from the body.

Gathering Wood: Women pick up wood from the ground by stooping from the waist, knees straight.[23] Sticks are sometimes broken by hitting them against a log on the ground.

Chopping Wood: The majority of both men and women who chop wood place the right hand up and the left hand down on the axe handle, swinging the axe over the right shoulder. One man slid his right hand on the handle as he chopped, up on the up swing and down on the down stroke. The women did not do this.

Kneading Bread Dough:[24] Dough is mixed by squeezing it with the

---

[23] Reichard, 1939, *op. cit.*, opposite pages 102 and 110.

[24] Bailey in *Navaho Foods and Cooking Methods* (American Anthropologist, Vol. 42, April, 1940), pages 270–290 lists a number of motor habits connected with cooking procedures. There fore only those not previously described will be mentioned here.

fingers, knuckles up, then kneading and pushing it down with the knuckles. The back of the hand is kept away from the body. A screwing motion (an outward rotation of the forearm) is combined with the pushing and squeezing. One woman states that dough must be kneaded very hard or it will be sticky.

One informant pulled the dough up from the bottom and far side of the pan, using her fingers. Then she pushed it down in the center squeezing it between her fingers as she clenched her hands on the near side of the pan. Her hands were used alternately, one crossing over the other.

Another woman kneaded in this same manner but used both hands together instead of alternately.

Using a Rolling Pin: Dough is kneaded into a three inch ball and placed on a low table which is in reality a cutting and kneading board. The dough is rolled with the right hand, as in rolling pottery coils, using a pushing and pulling motion with the palm down and a slight pressure on the roll. When the ball becomes an elongated roll, a rolling pin without handles is used to round the dough into flat cakes. Both hands are placed flat on top of the rolling pin, palms down, and the push-pull motion continues. Dough is turned around several times with the right hand to insure a circular cake.

One young woman used the same type of equipment described above but kneaded in the following manner: after cleaning the board by rolling and rubbing a small piece of dough over it she twisted a handful of dough from the main portion and kneaded it with the right hand for a minute or so. She pushed it away from her body by a pressure on the heel of her hand and pulled it toward her with her fingers. This is the same motion that white women use in kneading and it may have been learned in school. Then taking the rolling pin she flattened the dough, turning it around frequently, till it reached a diameter of eight inches and a thickness of one-half inch. She placed palms down on the pin and rolled it with her fingers extended forward, the palms and heels of her hands doing the work. Before placing the tortilla on the griddle she threw it back and forth several times from one palm to the other. She remarked that the rolling pin and board were used because she had gone to school and had not learned to shape tortillas by hand in traditional fashion.

Kneading Tortillas: When tortillas are shaped with the hands instead of a rolling pin, the three inch ball of dough is pinched and pulled into shape by using the fingers under and the heels of the hands on top of the patty. The flat cake is turned clockwise in the hands as it is shaped in this manner. When it enlarges it is slapped back and forth from one hand to the other between the palms, or between the partially flexed knuckles and the heels of the hands. The slapping motion is alternated with the pinching and pulling until the desired shape and size is obtained. When using the flexed knuckles the greatest force comes on the heel of the left hand which flattens the cake in the center.

Turning Tortillas: The unbaked tortillas are placed on the griddle by grasping the cake with the thumbs toward the body, fingers pointing down and on the far side of the cake which dangles from them. With a swing the piece is flipped away from the body and allowed to descend to the griddle, touching the far side first and being laid on from rear to front. The thumbs and fingers are removed just before the front edge touches the griddle. When they are baked on one side they are lifted from the front edge, fingers on top, thumb under, and turned over sidewards with an outward rotation of the arm.

Rolling Corn Dumplings: One informant showed how to roll two corn dumplings at once, one between the fingers and the other between the palms of her hands. She used a counterclockwise, circular motion of the right hand on the left.

Rolling Wheat Noodles: After thoroughly kneading wheat flour bread as described, noodles are shaped by rolling pieces between the palms into long coils like pottery coils.[25] The palms are facing, thumbs up, and the dough hangs down between the little fingers until a coil is made. It is then broken into short pieces and boiled.

Peeling or Cutting with a Knife: One informant states that some people, especially the older ones, avoid putting the point of a knife into any food such as a melon or meat. They connect it with war and believe a bullet will strike one in consequence. Older people who have been shot or wounded do not observe this restriction, but will freely put the point of a knife into food. Some young people avoid this act and some do not.

When using a knife, it is held in the right hand and the food to be cut in the left hand. The peeling or slicing is done away from the body as if whittling. Only one woman was observed cutting toward the body. When the investigator peeled potatoes with a group of Navaho women, she did twice as many in the same amount of time, using the knife in the right hand and peeling toward the body with the thumb as a lever. The Navaho women laughed at her method of handling the knife, but conceded that it was faster.

In using a modern potato peeler the Navaho woman also makes the whittling motions away from her body.

---

[25] Harry Tschopik, Jr. in his monograph *Navaho Pottery Making* (Peabody Museum Papers, Vol. XVII, No. 1, Cambridge, Mass., 1941) displays this coiling in a photograph, Plate III, b.

Meat is cut from the bones in like manner.[26] The bone is held in the left hand, the knife in the right, and the meat whittled off in irregular chunks. The bone is then cracked by several sharp blows with the knife and broken with the hands. One woman was seen cutting the meat around the bone lengthwise to make a long, flat piece. She held the meat in the left palm and cut toward her and toward the bone until she had a flat piece with the bone attached to one end.

Dusting the Hands: To dust powder (flour, sand, etc.) from the hands, the flexed fingers are flipped quickly across the thumb of the same hand in an explosive extension. This is repeated two or three times in quick succession. The fingers point towards the floor, the knuckles are up in the movement. Five women were observed doing this when cooking and three men while sandpainting.

Lifting Hot Coffee Pot: A pot is lifted from the fire with the hands in the reverse position from that taken by white women when lifting a pot. The thumb is pointed down and the fingers curled in around the handle. The arm is extended and rotated slightly outward while the pot is lifted and carried at arm's length down at the side. The woman bends from the waist, knees straight, to lift the pot.[27] The skirts are held back with the left hand.

Posture while Cooking: Most women sit, either on the ground or on a chair, while preparing as well as while cooking food. Four exceptions have been seen. One woman stood to knead bread, one to cut meat, and two to peel potatoes. In the latter case, however, the cook-shade was crowded and there was no room to sit. This habit seems to be a matter of etiquette as women are not supposed to remain standing in the hogan.

Grinding Corn: Corn is ground by pressure of a stone mano upon a stone metate. The metate is supported under the end nearest the grinder's body so that the stone slants down and away from her. She kneels back of the supported end, grasps the mano with both hands, knuckles up, fingers away from the body. Keeping the arms straight she swings back and forth with the knees as a fulcrum and produces a push-pull motion on the metate. The back is held parallel with the ground in this operation. The heaviest pressure comes on the push. As the corn falls off the stone on to the sheepskin which has been placed underneath for protection, it is scooped up with the left hand (the right continuing the grinding motion) and placed at the top of the stone. Then both hands are used again until more flour has fallen when the action is repeated. Corn is ground three times. The long end of a

---

[26] Reichard, 1939, *op. cit.*, photographs opposite pages 31 and 102.
[27] Reichard, 1939, *op. cit.*, opposite page 110.

grass brush is used to sweep the stone when finished. It is generally grasped with the little finger down toward the longer end and held nearly perpendicular while sweeping.

Grinding Salt: Mano and metate are used in the same manner as for corn grinding except that the heaviest pressure is on the pull rather than on the push stroke. The stone is swept with the curved hand, little finger down, to scrape the salt into a container.[28]

Stirring with "Corn Stirring Sticks": The sticks are held in the right hand like a pencil, fingers turned down and pointing toward the body, the pressure being applied against the last three fingers. Corn mush is lifted and turned with a twist of the wrist and arm. Some women hold the sticks with the right hand below the string that fastens them together (two-thirds of the way up) and the left hand above the strings to act as a fulcrum in stirring stick mush.

Bracing Pot Which Is Too Hot to Handle While Stirring: A strip of metal, eighteen inches long and two inches wide, is held in the left hand, one end placed against the rim of the hot container, and the other end resting on the ground braced under the left knee. The sitting position is with legs folded under and back to the left. This leaves the left leg on top to exert what pressure is needed to hold the brace in position. The right hand uses the corn stirring sticks in the usual way.

Sweeping: Women handle an American broom with right hand up and left hand down the handle, or vice versa, according to individual habit. Water is first sprinkled on the dirt floor to lay the dust by taking handfulls from a bucket or pan and scattering drops with an outward swing of the right hand. The sweeepings are directed into the depression under the stove or into the firepit. They are later scooped up with a flat piece of metal or cardboard and deposited outside. One woman states that to sweep the floor while guests are present would be a breach of etiquette.

Women sometimes use a Navaho broom (twigs and branches tied together on a handle) to sweep the "shade" and the ground around an outside fire. This broom is handled as we handle a bamboo rake when raking leaves. The sweepings are pulled toward the individual who moves along in front of them. Usually they are collected in the firepit and burned.

Sewing: Women sew both by hand and with a sewing machine. In both cases the movements are very similar to those of a good seamstress in white

---

[28] This difference in handling the mano and metate, and methods of brushing the stone may not be due to a difference in materials ground but to an individual variation. The investigator regrets that not enough cases of salt grinding have been observed to merit a generalization.

culture, except that no thimble has been observed. Threading a needle in one case was done from left to right, needle held in the left hand and pushed onto the thread held in the right. Turning a hem and pressing it in place was done by drawing the material across the edge of the sewing machine. The only awkward action was the use of a three yard gathering thread which tangled badly.

Most women sit on the floor while sewing, with legs folded under. One woman sat on a low box covered with a goatskin. While using the machine, the seamstress was seated in front, feet parallel on the treadle. She worked smoothly and efficiently. Paper patterns have not been seen. The women cut free-hand.

Softening Moccasins: Leather for moccasins is softened by grasping it with both hands and squeezing, rubbing, and kneading it back and forth between the fingers and knuckles. When one spot is completed, another is begun.

Repairing Tools: One old man was observed holding an axe in a very peculiar way when repairing the handle. He sat on the floor, knees out, legs flexed, and soles of feet together. Between the soles of his feet he braced the axe, using one foot slightly on top of the other. This left both hands free for his work.

Weaving: Since Reichard in her publications[29] has so adequately covered the subject of weaving, no discussion of motor habits will be made here.

Silversmithing: John Adair has a monograph on silversmithing in preparation.

GAME HABITS

Ball Throwing and Catching: On three occasions children have been observed throwing and catching a ten inch rubber ball. Two boys were about ten years old, one was seven. Each of them threw easily and well, using an over-arm shoulder pass, and getting considerable distance in the throw. It was noticeable, however, that their catching was totally inadequate and very awkward. The arms were stiff, no skill was evidenced in controlling the bounce, and in the entire time of playing *no one caught the ball.* Several times the seven year old clutched the ball to his chest, as would a much younger child in our culture, but dropped it again. It is possible that had the investigator been able to watch school children, the percentage of success would have been higher.

Motions Involved in Stick Dice Game: In this game three short sticks are thrown against a rock to bounce them in the air. As they fall on the ground

---

[29] Reichard, 1936, *op. cit.*

they are scored as counters. The motion used to bounce them on the rock is difficult to learn. They must be held firmly between thumb and fingers of the right hand (fingers pointing toward the left), or clutched in the hand with the fingers wrapped around them. The lower ends are made level by tapping them in the left palm. The throw is started about two feet above the stone and a quick, staccato thrust is made downward. The sticks are released a few inches above the stone. They hit and rebound into the air to fall on the ground. Without the hard thrust and quick release they do not rebound but scatter over the stone. Differences in strength of throw have been noted. One old woman made a violent movement which caused the sticks to bounce high and scatter much farther than normal. She was very interested in the game, determined to win, and was the cause of amusement because of her exaggerated motions in playing.

Certain "good luck" motor habits are seen in connection with this game. These are supposed to change bad luck to good, or to assure the thrower a "ten count" in the game and consequently a second throw. The three sticks may be held together and rubbed two or three times across the central stone with a quick, pulling motion toward the body. Sometimes two sticks are held in the left hand, and one stick in the right hand. The right hand stick is then struck sharply across the two left hand sticks and all three are dropped simultaneously to the ground. This gives the appearance of breaking the sticks, and is used to break bad luck or insure further good luck.

Archery: A social event during ceremonials is the shooting of arrows at a mark on the ground. The distances average forty yards. The bow is held in a cross-bow position rather than upright as in English shooting. The arrow is grasped at the nock end with a pinch grip of the thumb and bent forefinger. The third finger catches the string and the draw is made by a combined pull on arrow and string. This draw has been classified by Morse as the "Secondary or Assisted Pinch-Draw" and is illustrated by Lambert.[30] The arrow is released on the draw with no aim except an almost instinctive one. The bow is about two-thirds drawn and the release is made slightly above waist level. Only one man was seen using a grip which approximated the English three finger draw. As the arrow is released, the left hand raises the bow with a slight jerk which produces a bit more lift for the arrow in flight than it would ordinarily have at that angle. The average accuracy is about twelve to fifteen inches from the target. A few times the arrows have been seen to fall within three inches of the mark.

String Figures:[31] The making of string figures is infrequently done in the

---

[30] A. W. Lambert, Jr., *Modern Archery* (New York, 1929), p. 200, fig. 50.

[31] A. C. Haddon, *A Few American String Figures and Tricks* (American Anthropologist, Vol. 5, 1903), pages 213–224, discusses Navaho string figures.

summer because of its connection with lightning. However a few women, two men, and one child of two years have been observed making these figures. The start is made with the hands lifted in front of the chest, palms facing, fingers up. The string is threaded onto the upraised fingers in the same manner that white children thread the string for "cat's cradle." From there the process is different for only one person manipulates the string, producing a variety of complicated patterns all of which have names. Over twenty have been seen by the author. The fingers move with delicate picks and thrusts, lifting and dropping the strings rapidly into patterns. They move far too fast for the eye to follow the separate movements. The only suggestion that the author can make of the visual effect produced is that it would resemble a woman tatting, if instead of one string, she manipulated several at once in the same manner.

CEREMONIAL MOTOR HABITS[32]

Shaking the Rattle: The rattle is held in either left or right hand with the thumb on top of the handle and the rattle in line with the hand and forearm which is held at a right angle with the upper arm. The wrist is straight and the upper arm hangs loosely in a natural position from the shoulder close to the side. The head of the rattle is at the center line of the body.

The motion is an accented, short thrust downward of about two or three inches made with the forearm and wrist alone, held rather stiffly. The thrust is followed by an almost indistinguishable second thrust, up and down, making a sequence of long-short, long-short, in rhythm. At the end of the second down thrust, the hand is returned to the original up position. At the end of a song, or an interruption in the regular rhythm, the thrust is also changed and the rhythm broken to fit the song.

Individuals vary the length of the thrust and some do not always return the rattle to the up position each time, but make a pattern of their own with the long and short thrusts.

Hand Position of Singer: During a song the left hand of the Singer is placed on his left cheek, the heel of the hand at the chin bone, the little finger along the side of the mouth to a point below the left eye, and the first finger toward the lobe of the ear. The hand is slightly cupped. Often the eyes are closed at this time. In answer to a query as to the reasons for this hand position, one informant stated that a man places his hand thus if he starts a song to indicate to those assembled that he is leading that particular song.

---

[32] Mention of ceremonial sitting position has been made elsewhere in this paper. Only two other characteristic gestures will be described here, as Kluckhohn and Wyman in their publication on ceremonials have described minutely certain motions peculiar to specific ceremonials.

The rattle is held in the right hand as described above, if it is being used. Another informant stated that placing the hand in this position produced a better tone in the singer's voice. He gave no other reason.

TRANSMITTING MOTOR HABITS

Children have been observed learning motor habits in two ways: by unconscious imitation, and by the definite effort of an adult to teach them a specific skill. One informant says that she is teaching her seven year old daughter to weave. She has already shown her how to card and she does it fairly well. She adds that some girls don't need to be taught because they learn by watching, but her daughter had to be taught. One man said that when boys are old enought to handle the equipment properly, they are taught silversmithing and moccasin making.

A two year old boy was being taught by his father to shake a rattle in this manner. The man crossed his legs, tailor-fashion, and placed the boy in his lap in the same position. Reaching around him, he put the rattle in the boy's hand in the proper position and made the shaking movement several times with the boy's hand. He sang as he did this. At the same time across the room the boy's grandmother made shaking motions calling to him to imitate them. After much encouragement he finally did shake the rattle a few times by himself, and was praised.

A two year old girl was being taught to grind flour by an older sister, who placed the child in front of her at the metate with the child's hands between her own on the mano. As the older girl ground, the child's hand moved with hers.

Frequently a child is given a piece of dough to play with while the woman is kneading the bread. The child tries to imitate the motions of the woman in shaping the tortilla. Usually there is some slight evidence of success, but the attempt ends by the child eating the dough raw, or throwing it away.

A woman silversmith was seen teaching her eighteen month old son to hold a hammer. She placed him in front of the anvil in the proper position, put the hammer in his hand, and directed his hand to hit a point on the anvil. When his hand was released, he was remarkably successful in hammering on the same point alone.

A Singer held his son (less than two years old) in his lap and sang two notes to him over and over in the peculiar falsetto voice a Navaho sometimes uses. He made an effort to get the boy to repeat them after him and was successful only once. The child hit the exact notes that time, but either did not respond or got the wrong notes most of the time.

The same child was at another time placed in the driver's lap in an automobile and encouraged to steer the car with his hands between those of the driver. He enjoyed this so much that he cried when he was removed.

At another time he was sitting in his mother's lap when he saw cigarette smoke curling up toward him. Suddenly he reached for the smoke, pulling it toward him with his cupped hands and then applied the hands to his head and face. This was an obviously remembered action which he had at some time seen in a fumigation ceremony used by adults. It created quite a bit of excitement at the time as it was considered an unusual thing for a child of eighteen months to do. The investigator noted the action to be a remarkable evidence of delayed imitation.

While watching a game of Stick Dice, one small child picked up the sticks and threw them at the stone. He repeated this every time his turn came around. He never was successful in making them bounce as they should have done, but each time he straightened them carefully in his hand in imitation of the adults before he threw them.

RESTRICTIONS

Picking Up Objects: One informant volunteered information concerning picking up any object from the ground. He said that a man who knew how to do Eagle Catching was required to pick up all small objects (like matches) between the inner sides of his index and second fingers, palm toward the ground and fingers extended rather stiffly. Other people picked up such objects in the normal way (i.e. between the thumb and index finger).

Stepping over Objects: Several informants have stated that a menstruating woman must on no account step over a man or the man will be injured, even to the extent of becoming pregnant.

Hill[33] offers the information that no one must step over another person at any time, especially if wearing the token of a ceremonial which has been performed for him. This would bewitch the person stepped over. He also states that no one must step over the fire in the hogan as that is reserved for ceremonial ritual.

Turning Sunwise: In ceremonial procedure, whenever a turn is made it must be made in a sunwise or clockwise direction. The author, who did not at first know this, was repeatedly instructed to turn sunwise while participating in an Evil Way ceremonial. Later one informant mentioned turning sunwise as the proper procedure at any time.

Reversing Shoes: One informant stated that children are carefully watched lest they put their shoes on the opposite feet while dressing. If they

---

[33] W. W. Hill, field notes.

make this mistake, they must change them immediately to the correct feet, for only the dead wear their shoes reversed.

LEFT-HANDEDNESS

Hill[34] states that left-handedness appears to be more prevalent among Navahos than among whites. Some men prefer not to marry a woman who is left-handed because she is different. That it appears undesirable to some Navahos, is evidenced from the fact that several informants give information about attempts to change a child from left-handed to right-handed tendencies.

The father of one boy tied up his left hand in an effort to make him use the right hand instead. In this same family two other members of an older generation were left-handed, one of whom was successfully changed over and the other not.

One woman states that usually people don't care whether or not a child is left-handed, although the younger Indian parents try to correct it. Older ones don't mind. This would lead one to wonder if schooling had an influence on the pattern. The same informant states that a girl would be more likely to be corrected in a left-handed tendency than would a boy, because a girl couldn't cook and sew as other women do if she remained left-handed. She tried to correct left-handedness in one of her daughters by always handing things to her in her right hand, but decided it was too hard to do "because there is more strength in your left hand than in your right if you are left-handed."

TUSCAN SCHOOL
MAPLEWOOD, NEW JERSEY

---

[34] W. W. Hill, field notes.

World Distribution of Certain Postural Habits
GORDON W. HEWES

World Distribution of Certain Postural Habits*

GORDON W. HEWES
University of Colorado

FOR a long time it has been suspected that certain standing and sitting postures might be culturally significant (Mauss 1935). Although some (e.g., Driver 1937:118) have doubted that such behaviors are consistent enough to enable anyone to draw useful conclusions about them, others have not hesitated to deal in systematic fashion with the postural habits of particular cultures (Irizawa 1920–21; Mead and Macgregor 1951; Bailey 1942). For example, following a discussion of Mohave sitting postures, Kroeber (1925:728) asserted, "this [i.e., posture] is one of the most interesting matters in the whole range of customs. . . . " However, in spite of this interest, there have been almost no attempts to bring the scattered data together into a comprehensive world-wide framework. Mauss' *Les techniques du corps* (1935) was a stimulating program for such an undertaking—nothing more.

Human postural habits have anatomical and physiological limitations, but there are a great many choices, the determinants for which appear to be mostly cultural. The number of significantly different body attitudes capable of being maintained steadily is probably on the order of one thousand. Certain postures may occur in all cultures without exception, and may form a part of our basic hominid heritage. The upright stance with arms at the sides, or with hands clasped in the midline over the lower abdomen, certainly belongs in this category. A fourth of mankind habitually squats in a fashion very similar to the squatting position of the chimpanzee, and the rest of us might squat this way too if we were not trained to use other postures beyond infancy. Anthropoid postures may shed some light on the problem of which human ones are most likely to be "natural" or precultural, although ape limb proportions would deter us from relying too heavily on such evidence.

Of factors which affect postures, aside from the biological substrata, we might start with sex-differentiating conditions such as pregnancy and lactation, which possibly render certain sitting positions more frequent among adult females, and nutritional conditions (not wholly independent of culture), which may determine the amount of fat accumulated in posturally strategic parts of the body. Fear of genital exposure, whatever its etiology, seems to play an important role in postural customs (or at least in their rationale) in many cultures. Clothing and footgear, such as heavy boots and tight-fitting skirts, doubtless exert their effects on ways of sitting. Artificial supports—whether logs, rocks, stools, pillows, back rests, benches, or chairs—are highly significant, with complex histories and manifold cultural interconnections. There is also a relation of types of house-construction to posture.

* Read at the December, 1953, Annual Meeting of the American Anthropological Association, in Tucson, Arizona.

The influence of techniques and activities like textile weaving, fire making, wood carving, food grinding, playing of musical instruments, canoe paddling, or the use of gaming devices, all requiring the maintenance of particular bodily attitudes, cannot be denied. Nearly every new tool or machine must be adjusted to some body posture or sequence of postures. The push button represents the ultimate attenuation of environmental control through postural adjustments; vocal cord vibration to actuate servomechanisms, though still neuromuscular, cannot be described as postural.

Terrain and vegetation may influence out-of-door sitting or standing habits. In some regions the existence of high grass may force herdsmen to watch their flocks from a standing position, whereas in a short-grass or tundra region, herders may watch the stock while squatting or sitting down. In our own culture, moist, snowy, or muddy ground clearly inhibits sitting down, whereas a reasonably dry lawn may invite us to do so (Mauss 1935:280, 286).

Habitual excretory or burial postures may become tabooed in other situations. It is altogether possible that the rarity of the deep squat in our culture is due to this kind of repression.

Finally, several writers suggest that infant carrying customs may affect the postural maturational sequence, and the ease or difficulty with which children or adults can assume certain postures. If tight swaddling or cradling can influence later postural habits to anything like the degree that some authorities have claimed is the case with personality, the effects on sitting and standing behaviors should be indeed remarkable.

Some other ramifications of the cross-cultural study of postural habits can only be touched upon. Animal research on the infraprimate level seems unlikely to yield much of direct interest, since Magnus (1925), DeKleyn, and Sherrington have shown how the primitive postural reflexes are normally suppressed by cortical centers in the higher mammals. Bull (1951) and others have been working on the relation of reflex-related postural tensions and emotion, however, and there is a psychiatric interest in posture exemplified by the work of Schilder (1935), Feldenkrais (1949), and Quackenbos (1945). *Psychological Abstracts* contains about forty items dealing with the relationships of bodily posture to such diverse topics as visual acuity, lateral dominance, psychoanalysis, fatigue, and metabolic rate. Research on the metabolic efficiency of various postures has to date been restricted, because of our cultural traditions, to standing, chair-sitting, and recumbent positions (Tepper and Hellebrandt 1938; Larsen 1947).

The foregoing remarks suggest that anthropologists and others with opportunities to make cross-cultural observations of human postural habits could organize their data in terms of several levels of relevance. The first level is that of applied physical anthropology (King 1948; White 1952; Hertzberg and Daniels 1952) or "biotechnology." In recent years, the demands of machine and vehicle design have stimulated research in an interdisciplinary field where information on postural habits in different cultures might be very useful. Human engineers might profitably experiment with postural patterns borrowed from outside our own cultural tradition.

The second level of relevance is that of functional interrelations of postures and nonpostural cultural phenomena, discussed above in connection with techniques and activities, terrain and vegetation, clothing, status and role differences, etc. Are there cross-cultural regularities in these functional relationships? Some of the data to follow suggest that there are.

A third level concerns psychological and psychiatric implications of postural behaviors. Findings in this area would of course feed back to the applied anthropology and physiology level.

A fourth level of relevance is culture-historical, along the lines indicated by Boas (1933) in a discussion of criteria for historical reconstruction: "Certain motor habits . . . may be stable over long periods." It is possible that postural habits could be used like other culture elements in reconstructing past diffusions and contacts. The Samoan occurrence of a special cross-legged position (Fig. 1, 83) which is elsewhere apparently linked with religious diffusions from India into Southeast and Eastern Asia may be a case in point. More precise determinations of the distributions of various postures might reveal definite blocs of postural tradition which could have deep culture-historical meaningfulness.

A fifth and final level of interest is phylogenetic. The role of postures and their functional interrelations with environment (cultural as well as noncultural) seem to have been important. The development of the upright stance and of bipedal locomotion took place well before the simian brain case and face had been modified into essentially human features (Washburn 1951), which suggests that standing, sitting, squatting, and recumbent postures preceded the emergence of fully human behavior. Here efforts could be made to bring together the results of studies of anthropoid ape postural behaviors, including those on maturation, on limb bones and joints of the pre- and protohominids, and theoretical syntheses such as Orione's (1950), on the interconnections of eye, hand, posture, and symbol-using capacities in mammalian evolution.

For the following discussion, I have recorded information on about one hundred of the commonest postures, chiefly of the sitting, kneeling, crouching, and squatting varieties. Only about ten standing positions are included; many upright stances seem to be human universals, for which reason I soon ceased to record information on them. Sleeping or reclining postures have been entirely omitted, owing to the paucity of data (Peter 1953). Maturational aspects of posture and the pathology of posture (as in Parkinsonism or catatonic schizophrenia) have not been considered here. Dance positions have also been excluded, perhaps arbitrarily; for the most part they are held for only a few seconds at a time, whereas the attitudes dealt with here are maintained for minutes, at least. Finger and hand gestures, ways of holding the head, facial expressions, the tone of the abdominal musculature, and many other interesting aspects of posture have also had to be left out of this paper.

Written descriptions of postures are rare in ethnographic literature, and when they occur they are often so ambiguous as to be almost worthless unless supported by photographs or drawings. There have been a few studies, like

Bailey's (1942) on Navaho motor habits, Mead and Macgregor's Balinese growth research (1951), Kano and Segawa's thorough photographic presentation of the Yami of Botel Tobago (1945), and Irizawa's monograph on Asiatic sitting postures (1920-21) which do provide excellent material on our subject. The University of California culture element surveys of the 1930's, covering a large part of western North America, contain perhaps the most complete record of postures on a cross-cultural basis, even though some of the lists lack postural items and others are worded so vaguely as to be unusable. The Human Relations Area Files, Inc., in its ambitious program, has not overlooked postures in its *Outline of Cultural Materials* (Murdock and others 1950:69), but that portion of the files now on deposit at the University of Colorado Library yielded information of very uneven quality. The dearth of postural data in the HRAF is, however, merely a reflection of the inattention of ethnographers to postural behavior generally. The otherwise comprehensive *Notes and Queries on Anthropology* (Royal Anthropological Institute 1951) overlooks the possibilities of observation of or inquiry into postures.

By default, therefore, I have been forced to rely chiefly upon published photographs from a great variety of sources—ethnographies, books of exploration, and especially popular geographical magazine articles—supplemented by occasional descriptive statements. Some posture information has been obtained in this manner from about 480 different cultures or cultural subgroups, distributed as evenly as the available sources permitted. Thirty-four of the cultures surveyed are ancient or known only from archeology, where postural data consist of figurines, carvings, or paintings sufficiently realistic to be classified. Obviously there is a different sampling problem with postures represented in art than there is with those noted in ethnographic reports or contemporary photographs. Nevertheless, from some ancient cultures, such as Ancient Egypt, the information appears to be more complete than from some supposedly well-known recent peoples.

In using photographs, one must assume that it is possible to distinguish between postures imposed upon the subjects by the photographer, and those which are habitual or indigenous. There is little strain on this assumption when it comes to separating pictures of individuals engaged in routine tasks, or resting informally in or about their own dwellings, from stiffly posed studio-type shots, against painted backdrops. There are, however, pictures in which the subjects have certainly been arranged in a line for purposes of photographic composition, but in which seemingly indigenous postures occur. For example, in the only two group photographs of the now extinct Tasmanians, fortunately taken out of doors, there are some individuals seated on the ground in what was probably a habitual Tasmanian posture, even though others in the same photographs were sitting rigidly on chairs. The warpage in the data arising from the preponderance of males in ethnographic and travelers' photographs reflects several factors. Some cultures, as in much of the Islamic world, keep women out of sight of strangers, or at least discourage their being photographed. For the world as a whole, perhaps, males are more likely to be encountered

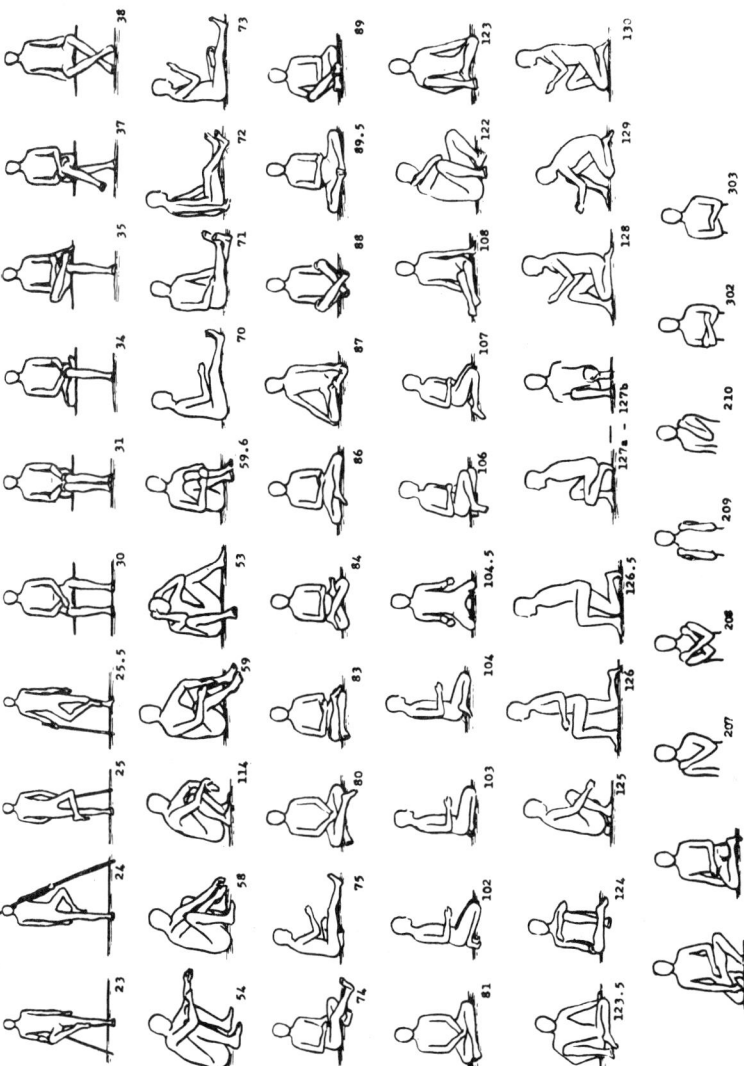

FIG. 1. A portion of the postural typology used in the compilation of data for this paper. Drawings are for the most part based on photographs in the ethnographic literature. Head and arm positions, unless stated otherwise in the accompanying discussion, are not typologically significant. No. 23, for example, could be standing with his left hand on his hip, or resting it on his left shoulder, and his standing posture would be considered the same for present purposes.

out of doors, in the direct sunlight, and hence are more likely to be accessible to photographers without special film or lighting equipment.

One check on the reliability of photographic postural data lies in a comparison of pictures taken among the same group or within the same culture by two or more ethnographers or other visitors, at different dates. Another check exists in the much less frequent case where a written account of postural habits can be compared to a set of photographs.

Having determined that written descriptions of postures are less efficient than sketches of figures, I began this preliminary survey by recording all postures encountered in a large, culturally random sample of ethnographic photographs, by means of small drawings alongside which I wrote notes on the ethnic group, sex, and approximate age of the individuals shown, and the activity (or inactivity) indicated. Eventually several hundred sketches, with notes, were assembled, and from them about one hundred postures were selected for a typology-sheet which was used instead of sketches to facilitate through the use of arbitrarily assigned type-numbers the recording of larger masses of information. The same system was used to code the relatively few written descriptions of postures found in the literature. A tabulation sheet, with the names of cultures or tribes along the Y-axis and the posture-type numbers along the X-axis, was then prepared, with provision for an indication of the frequencies and sex-associations of the postural data tallied in each cell. Thirteen world distribution maps were next compiled from the tabulation, one of which simply shows the location of groups, ancient or modern, from which one or more usable data were obtained. Space considerations preclude the reproduction here of all the maps; two of the most significant have therefore been selected. A partial reproduction of the posture typology (Fig. 1) may be helpful to future investigators of standing and sitting habits. North-central Siberian cultures are not represented because of a dearth of illustrations in the small number of sources on that area available to me. Eastern North America is inadequately represented primarily because its native peoples were displaced, where they were not destroyed, prior to the invention of photography; their descendants have adopted, along with European clothing, house types, furniture, and tools, the postural habits of Western civilization.

Inclusion of the complete bibliography of sources utilized for the tabulation and mapping of postural traits would consume too much space also, since it runs to 212 titles.

Nilotic one-legged resting stance (Fig. 1:23–25.5)

The "classic" *Nilotenstellung* occurs not only among the Shilluk and their neighbors in the southern Sudan, but in Nigeria (Elkin and Fagg 1953), Iran (Singer and Baldridge 1936), India (Koppers 1944), Ceylon (Buschan 1923), Australia (Elkin and Fagg 1953), in South America among the Nambicuara and Yecuana (Steward 1948), and, if we can accept the California element-survey data, rather widely in the American Southwest (Gifford 1940; Stewart

1942). The Ceylon instance may be questionable, since the individual is shown leaning against a tree, and the sole of the bent leg is not in contact with the side of the opposite leg; a similar objection can be raised against a Santa Cruz Island (Melanesia) example. All instances of this stance seem to be represented by males. Aside from the Nilotic Sudan, where this posture is known to be assumed by cattle herders, we have little information on its cultural functions (Hewes 1953). Gifford notes that the men of Walpi pueblo rest this way in the fields while hoeing.

A series of stances at least conceptually relatable to the *Nilotenstellung* can be recognized, in which the feet are crossed, with one foot flat and only the toes of the other touching the ground, or crossed in any case so that the main weight of the body is on one leg. Plotting the distribution of these lesser one-legged resting stances, we find at least a tendency for them to form clusters with the genuine Nilotic posture or positions very close to it, in Africa, Australia, and Melanesia.

Chair-sitting postures (Fig. 1:30–38)

Eleven common ways of sitting on chairs or chair-high benches or stools have been recorded and their distributions mapped. Although such postures are probably far more frequent in cultures where chairs and benches are common articles of furniture, such furniture is certainly not a requirement. People can sit much as we sit on chairs or benches wherever they have access to suitable large logs, boulders, cut-banks, or the edges of house platforms, stone or earthen terraces. Many Melanesians and Papuans manage to sit as we do, using the edges of house platforms. Back support may increase the restfulness of the sitting posture, but seems to be a minor factor. Even though floor- or ground-sitting or squatting may be the rule in cultures like those of Bali or Japan, people often sit on ledges and the like. The only large area in which no pictures of chair-sitting postures (other than obvious nineteenth-century studio portraits) were encountered was Australia.

Sitting on chairs, benches, or comparable supports ranges from the stiff, formal posture with trunk vertical, feet and legs held together along the midline, with hands resting on the upper thighs or clasped in the lap, to a variety of more relaxed positions with legs spread apart, sometimes asymmetrically, or with the legs crossed at the knees, midshins, or ankles, etc. It is very likely that in a number of cultures, the photographic situation is perceived as one calling for the most formal of postures, so that the rigid, legs-together position is commonly elicited. Historical changes in the etiquette of sitting postures can be followed quite well in European art prior to the rise of photography, as well as in the little more than a century since its invention.

Chairs and benches go back at least 5,000 years in Egypt and Sumer. While rulers and magnates (along with their divine counterparts) monopolized the chair, Egyptian commoners—artisans and servants—are depicted sitting on chair-high benches in the early dynastic period. Space is lacking for an extended discussion of the history of chair-using and its social-status connections. Chairs came relatively late into Chinese culture, and never really overcame the floor-sitting postures traditional in Korea and Japan. The New World high civilizations preferred low stools, on the whole, and these only for persons of high rank, though true chair-high seats developed in a few areas such as Manabí, Ecuador. The use of stools and other low sitting supports is quite widespread in Africa and the tropical forests of South America, and has a sporadic distribution in Oceania and North America; they tend to be reserved for males, and often only for men of authority.

Deep squatting postures (Fig. 1:54, 58, 114)

Squatting with the soles of the feet flat and the buttocks either actually resting on the ground or floor, or only an inch or two above it, has a very wide distribution except for European and Europe-derived cultures. Absence of reports of the deep squat from northern Asia probably can be attributed to the general weakness of my data on that area. The only European examples I found were ancient—on a Greek metal vessel and on a LaTène Iron Age vessel based on the same motif; significantly, the squatting figure represented a daimonic individual playing a Pan's pipe, which suggests that the original Greek artist regarded the posture as uncouth or primitive. In the culture of the contemporary United States, the deep squatting posture is reliably reported to occur among males in the backward mountain communities of the southern Appalachians and the Ozarks. In the world at large, women seldom use the deep squat, at least in public or in situations likely to be photographed. The role of taboos against female genital exposure in determination of acceptable or nonacceptable feminine postures is presumably important, but the evidence is slight. The deep squatting posture is of course a near-universal defecation position outside of a few occidental cultures, and is often used when urinating, sometimes by one, sometimes by both sexes (Ellis 1911:66).

A posture resembling the squat, but in which the legs are only partly flexed (Fig. 1:53, 59, 59.6), has nearly world-wide distribution, again chiefly among males. It is frequently assumed by males (and by trousered females) in our own culture when the customary chairs or benches are unavailable.

Sitting with legs stretched out (Fig. 1:70–72)

Sitting on the ground or floor with the legs stretched out in the midline, sometimes with the ankles or knees crossed, is one of the best cases of a feminine postural habit. Despite the heavy preponderance of information concerning males in our data as a whole, there are very few male instances of this position. It is associated with a number of tasks commonly assigned to women in the division of labor, such as the use of belt or horizontal looms, or the weaving of mats. Such a posture may also be well suited to the nursing of infants, especially if the mother wishes to engage in other activities simultaneously. Melanesian women often seem to rest in this position, while their menfolk sit about in deep squats or with the legs partly flexed (Fig. 1:59).

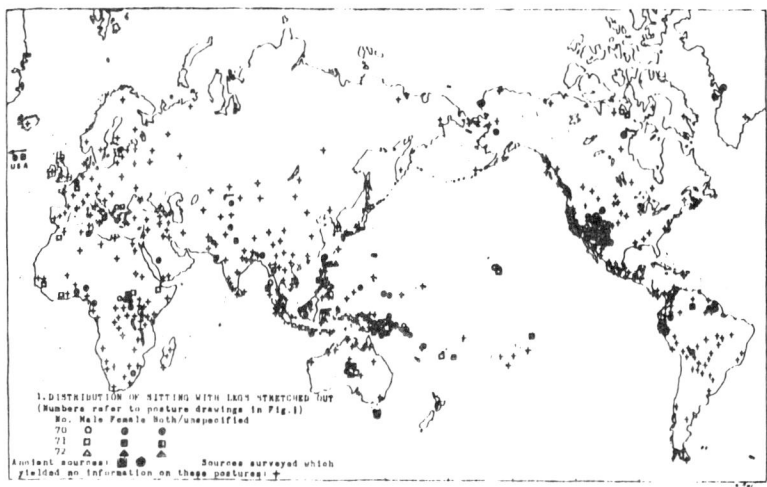

There are several similar postures in which one leg is stretched out and the other is variously flexed so that the foot lies above the opposite knee, beneath it, or is sat upon (Fig. 1:73–75). Unlike the legs-stretched-out position, this is not a particularly feminine habit, and little of significance can be made out from its distribution, at least in the present state of our information.

Cross-legged or "tailor-fashion" postures (Fig. 1:80–89.5)

Sitting with the legs crossed, in what we loosely call in English "tailor-fashion" or "Turk fashion," actually breaks down into some eight or ten dis-

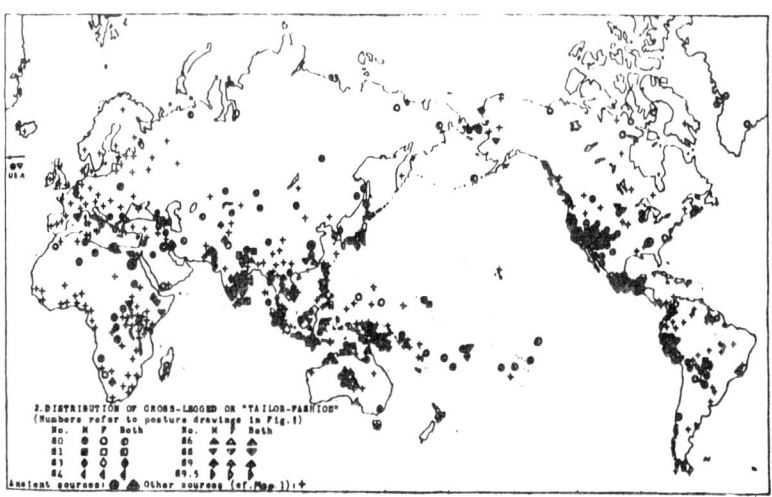

tinct postures. The basic cross-legged position is very widespread, although it is probably not a human universal trait. Kroeber once described this as the commonest sitting posture, though if the count were to be made by individuals rather than by separate cultural entities, chair-sitting (because of its high frequencies in Euroamerican and Chinese cultures, and because the deep squat is the posture preferred for daily use by the Indian and Southeast Asian masses) might take first place. The specialized variants of the cross-legged position cluster in India and Southeast Asia–Indonesia, where they have well-known religious implications. In our own culture, informal observations suggest that the basic cross-legged posture is assumed somewhat less readily than that with the legs partly flexed (Fig. 1:59, 53) by males and trousered females.

The religious diffusion of some of the cross-legged postures from India with Hinduism and Buddhism seems to be unquestionable; thus in Japan one finds the *hanka-fuza* (Fig. 1:83) and *kekka-fuza* (Fig. 1:84) practically restricted to the Zen Buddhists. It is therefore of considerable interest to find the *hanka-fuza* position as a well-established Samoan custom—apparently uniquely in Polynesia. Sitting cross-legged so that the sole of one foot lies on the opposite thigh occurs in Samoa, Japan, Bali, Cambodia, Thailand, and in India—where it is one of the Yoga *āsanas* discussed by Coomaraswamy and Duggirala (1917). Such postures seem to be quite unknown in western Eurasia, Africa, and the Americas.

Kneeling on knees and feet, or knees and heels (Fig. 1:102–104.5)

These postures are fairly widespread, with a suggestion that they tend to be used mainly by females in much of Negro Africa, Mexico, and parts of Indonesia. Sitting in what might be described as a "deep kneel," with the feet under the buttocks, dorsal surfaces down, is the normal Japanese sitting posture for both sexes. In the Islamic world this is a regular prayer position, and perhaps is associated with supplication in the Eurasiatic higher civilizations generally, if not even more widely.

Sitting with the legs folded to the side (Fig. 1:106–108)

Like sitting with the legs stretched out, this is a strongly feminine custom, found in the American Plains and Southwest, Melanesia, South America, southern Africa, and—where there are no chairs or benches available—in Euroamerican culture. Rarely, it is a male habit, as among the Mohave (Kroeber 1925). Tubular skirts or similarly restricting clothing may be associated with this usage, but it occurs often enough among quite skirtless groups to make such an explanation inadequate. Exposure avoidance might provide a more plausible explanation, if this motivation can be given priority over a postural habit.

One knee up, other down and flexed (Fig. 1:122–131)

There are numerous variations of asymmetrical kneeling or sitting postures in which one leg is flexed with the knee up, and the other leg is bent down or to the side, one limit of which is reached when the knee and foot of this other leg are in contact with the ground or floor. Versions of these asymmetrical positions are popular among the Australian aborigines, common in Africa and much of North America, but seemingly rare in India, South America, Europe, and Polynesia. They are very seldom used by females. The asymmetrical kneeling or genuflection known in the American West as the "cowboy squat" is well represented in Classical Greek sculpture for both sexes, occurs in Ancient Egypt, native North America, and scatteringly elsewhere. Kneeling positions of this type have frequently been used for archery, and examples have been noted from Spanish Mesolithic cave-paintings all the way to Tierra del Fuego, by way of some archers carved on the Parthenon friezes. The fire-drill is often twirled by someone in virtually the same posture, which is favored likewise by the "crapshooter."

Many other postural habits were systematically noted, but in the materials sampled from 480 cultural groups, instances of any one were too few to warrant mapping. Standing with one or both hands on the hips proved to be a very common informal pose for persons of both sexes in a great many cultures. Although there is a weak tradition in our own culture that the way in which the hand is placed on the hips is sexually significant and that for a male to depart from masculine usage is a symptom of homosexuality, no consistent sex preference for hand position in these postures could be determined, and group photographs showed clearly that there is considerable variation in this habit within almost any given culture. Hand-on-shoulder resting—sometimes with one hand on the same side, sometimes on the opposite shoulder, sometimes with both hands resting on the shoulders of their own sides, sometimes crossed over to the opposite shoulders—had a sporadic distribution (Fig. 1:207–210). This habit is absent or at least very rare in our culture, at any rate. On the other hand, folding the arms across the chest, either with one hand tucked under the opposite arm, or both tucked into the opposite armpits, is exceedingly common, and probably almost as frequent as standing with hands on hips (Fig. 1:302–303). Such an arm position is highly functional if one is trying to keep warm in a heavy blanket or robe which has not been provided with a fastening, but it often occurs also among naked or nearly naked people.

In spite of the tentative nature of my information, I think there is good reason to believe that many postures are not only culturally determined, but will exhibit the kinds of geographic distributions we have come to expect for other features of cultural behavior. While some human postures may be archetypical or precultural, others have probably been diffused like other items of culture. A note on the diffusion of a posture (the cross-legged sitting position) comes from Quain's Fiji study, in which he reported that older inland natives of Vanua Levu could recall when this fashion was introduced from Tonga (1948:19), first as a chiefly prerogative and only later as a custom of commoners.

The extent and vigor of postural etiquette evidently varies from one culture or area to another, with some societies going to great lengths to ensure postural propriety on all public occasions. Many cultures maintain careful distinctions in posture on the basis of sex, and there are others which emphasize age and status considerations in the manner of sitting or standing. Postural conformity is enforced as a rule by the same methods as conformity to other rules of etiquette—by ridicule (mentioned in several sources), verbal scolding, or by physical punishment where deviation from the postural norms verges on lese majesty or deliberate indignity to a superior. While our culture has perhaps relaxed its postural codes since the nineteenth century, certain areas of it preserve archaic postural etiquettes backed up by formidable sanctions, as in the military drill regulations. Postural deviations may also have a positive significance, as Blackwood mentions for the Buka Passage area, Northern Solomons, where a woman who sits with legs stretched out is regarded as openly inviting sexual intercourse (1935:125).

In our culture, the armament of science has been concentrated on normative efforts directed against a rather narrow band of our common postures, to the general neglect of many other postures which we use more frequently and with greater socioeconomic effect. Physiologists, anatomists, and orthopedists, to say nothing of specialists in physical education, have dealt exhaustively with a few "ideal" postures—principally the fairly rigid attention stance beloved of the drillmaster, and student's or stenographer's habits of sitting at desks. The English postural vocabulary is mediocre—a fact which in itself inhibits our thinking about posture. Quite the opposite is true of the languages of India, where the yoga system has developed an elaborate postural terminology and rationale, perhaps the world's richest.

In conclusion I should like to stress the deficiencies in our scientific concern with postural behavior, many of which arise simply from the all too common neglect (by nonanthropologists) of cross-cultural data.

REFERENCES CITED

(The writer wishes to thank Prof. Eiichiro Ishida of Tokyo, Dr. Frank Essene of the University of Kentucky, Dr. Carroll Riley of the University of North Carolina, and Dr. Beate Salz of Chicago for personal communications and references on the subject of postures.)

BAILEY, FLORENCE
    1942 Navaho motor habits. American Anthropologist 44, No. 2:210–34.
BLACKWOOD, BEATRICE
    1935 Both sides of Buka Passage. New York, Oxford University Press.
BOAS, FRANZ
    1933 Relationships between north-west America and north-east Asia. *In:* The American Aborigines, Their Origin and Antiquity, ed. Diamond Jenness, pp. 357–70. University of Toronto Press.
BULL, NINA
    1951 The attitude theory of emotion. Nervous and Mental Disease Monographs, No. 81. New York.
BUSCHAN, GEORG (ed.)
    1923–26 Illustrierte Völkerkunde. 3 vols. Stuttgart, Strecker und Schröder.

COOMARASWAMY, ANANDA and DUGGIRALA
  1917   The mirror of gesture. Cambridge, Mass.
DRIVER, HAROLD
  1937   Culture element distributions: VI, southern Sierra Nevada. University of California Anthropological Records 1, No. 2.
DUPOUY, WALTER
  1953   Algunos casos de postura nilótica (Nilotenstellung) entre indios de Venezuela. Boletín Indigenista Venezolano 1:287–97
ELKIN, A. P. (and WILLIAM FAGG)
  1953   The one-leg resting position in Australia. Man 53:64.
ELLIS, HAVELOCK
  1911   Man and woman. New York, Charles Scribner's Sons.
FELDENKRAIS, M.
  1949   Body and mature behavior; a study of anxiety, sex, gravitation and learning. New York, International University Press.
GIFFORD, E. W.
  1940   Culture element distributions: XII, Apache-Pueblo. University of California Anthropological Records 4, No. 1.
HERTZBERG, H. T. E. and GILBERT S. DANIELS
  1952   Air Force anthropology. American Journal of Physical Anthropology, n.s. 10, No. 2:201–08.
HEWES, GORDON W.
  1953   The one-leg resting position. Man 53:180.
IRIZAWA, TATSUKICHI
  1920–21   Nihonjin no suwari-kata ni tsuite. Shigaku-zasshi 31, No. 8.
KANO, TADAO and KŌKICHI SEGAWA
  1945   The illustrated ethnography of Formosan aborigines. Tokyo, Seikatsusha Ltd.
KING, BARRY G.
  1948   Measurements of man for making machinery. American Journal of Physical Anthropology, n.s. 6, No. 3:341–51.
KOPPERS, W.
  1944   India and dual organization. Acta Tropica 1:72–119.
KROEBER, A. L.
  1925   Handbook of the Indians of California. Bureau of American Ethnology, Bulletin 78.
LARSEN, ELEANOR W.
  1947   The fatigue of standing. American Journal of Physiology 150:109–21.
LINDBLOM, GERHARD
  1949   The one-leg resting position (Nilotenstellung), in Africa and elsewhere. Stockholm, Statens Etnografiska Museum.
MAGNUS, R.
  1925   Animal posture. Proceedings of the Royal Society 98B:339–553.
MAUSS, MARCEL
  1935   Les techniques du corps. Journal de psychologie normale et pathologique 32:271–93.
MEAD, MARGARET and FRANCES C. MACGREGOR
  1951   Growth and culture. New York, G. P. Putnam's Sons.
MURDOCK, G. P. and others
  1950   Outline of cultural materials. New Haven, Human Relations Area Files, Inc. 3d rev. ed.
ORIONE, JULIO
  1950   Teoria visual espacial. Buenos Aires, N. Zielony.
PETER, H. R. H., PRINCE OF GREECE AND DENMARK
  1953   Peculiar sleeping postures of the Tibetans. Man 53:145.

QUACKENBOS, H. M.
    1945 Archetype postures. Psychiatric Quarterly 19:589–91.
QUAIN, BUELL
    1948 Fijian village. Chicago, University of Chicago Press.
ROYAL ANTHROPOLOGICAL INSTITUTE OF GREAT BRITAIN AND IRELAND, COMMITTEE OF
    1951 Notes and queries on anthropology. London, Routledge and Kegan Paul Ltd. 6th ed.
SCHILDER, PAUL
    1935 The image and appearance of the human body. London, Paul, Trench, Trubner.
SINGER, CAROLINE and CYRUS LEROY BALDRIDGE
    1936 Half the world is Isfahan. New York, Oxford University Press.
STEWARD, JULIAN (ed.)
    1948 Handbook of South American Indians. Bureau of American Ethnology, Bulletin 143, Vol. 3.
STEWART, OMER C.
    1942 Culture element distributions: XVIII, Ute-Southern Paiute. University of California Anthropological Records 6, No. 4.
TEPPER, R. H. and F. A. HELLEBRANDT
    1938 The influence of the upright posture on the metabolic rate. American Journal of Physiology 122:563–68.
WASHBURN, S. L.
    1951 The new physical anthropology. Transactions of the New York Academy of Sciences, Ser. 2, 13, No. 7:298–304.
WHITE, ROBERT M.
    1952 Applied physical anthropology. American Journal of Physical Anthropology, n.s. 10, No. 2:193–99.

Panorama of
Dance Ethnology
by Gertrude Prokosch Kurath

Panorama of Dance Ethnology

by Gertrude Prokosch Kurath

APPROACH

DANCE as a reaction to life has a long tradition that encircles the globe. Dance ethnology, however, has come into being only within the last few decades. Though studies of dances are to date still individualistic and experimental, the literature as a whole is comprehensive enough so that the time is ripe for a co-ordination of the many different approaches.

COVERAGE AND GAPS

In the course of time, dances from probably every corner of the globe, as well as relevant customs now long vanished, have found their way into literature, for the most part in travelogues or sociological works. The literature of accurate description or analysis falls almost entirely within the last fifty years, and is now also respectable in quantity.

European dance ethnology received impetus from the research of Cecil Sharp in England, early in the twentieth century. Today all European countries can boast large, and sometimes systematic, government-sponsored collections of folk dances, particularly England and the Balkans. The names of the scholars who are most prominent in this work will appear often in the following pages. Wolfram further cites for Austria the work of Ilka Peter, Herbert Lager, and, as the "grand old man," Raimund Zoder; Hans von der Au and Felix Hoerburger, in Germany; Bianca Maria Galanti, in Italy; and Joan Amades and Aurelio Capmany in Catalonia.[1]

While the huge territory of the Union of Soviet Socialist Republics had seemed to be represented only by scattered reports of popular dances in anthologies, the ethnologist, I. I. Potekhin, sent (Dec. 12, 1958) a list of 139 bibliographical items which starts with the year 1848 and reaches to the present time. It includes anonymous surveys dealing with the dances of the U.S.S.R. (*Tantsi Narodov SSSR*), and many items containing choreographies and music of special regions, for instance, Azerbaijan (Almasadze 1930) and Moldavia (Onegina 1938). Accounts of Russian dances have been published in Berlin (Moiseyev 1951), Sophia (Okuneva 1951), Prague (Berdychova 1951), and Yakutsk, Siberia (Zhornitzkaya 1956). Further titles and an appraisal will appear in a 1960 issue of *Ethnomusicology*.

Labanotation, a technique of dance notation described below, under "Second Circuit," is in full swing in Europe. According to Knust, who terms it "Kinetography Laban,"

GERTRUDE PROKOSCH KURATH is Co-ordinator of the Dance Research Center in Ann Arbor, Michigan, U.S.A. She was born in 1903 and educated at Bryn Mawr College (M.A., 1928, History of Art) and at the Yale University School of Drama (1929-30), in the U.S. She also received extensive training in music practice and theory, and in several systems of art dance as well as folk dancing, in Germany and the U.S. From 1923 to 1946, she was active as teacher of Modern Dance, as concert performer, and as producer of pageants and dance dramas. Some of her choreographic compositions were based on research in European dances of the Middle Ages and Renaissance, and in American Indian and jazz dance. Since 1946, KURATH has concentrated on dance ethnology and ethnomusicology. Her research has included field work among the Aztec, Otomí, Tarascan, and Yaqui Indians of Mexico, and the Iroquois, Cherokee, Ottawa, Chippewa, Menomini, Fox, Tewa, and Keresan Indians of North America. Among her many articles and co-authored books, she wrote the dance entries for the *Dictionary of Folklore, Mythology and Legend*, and articles on dance for the *Encyclopedia Americana* and the *Encyclopaedia Britannica;* she is also dance consultant for *Webster's New International Dictionary*.

In June of 1958, KURATH accepted a suggestion from the Editor that she write a survey of dance ethnology for CURRENT ANTHROPOLOGY. To supplement her data on certain parts of the world, she embarked on eighteen months of correspondence with scholars in various countries. She received answers, information, ideas, manuscripts, reprints, or illustrations from the following correspondents: Renato Almeida, Henry R. Baldrey, Franziska Boas, Donald Brown, Richard L. Castner, Nadia Chilkovsky, Dance Notation Bureau of New York, Edward Dozier, Blanche Evan, William N. Fenton, Josefina Garcia, Erna Gunther, William Holm, Katrine A. Hooper, James H. Howard, Shirley W. Kaplan, Maud Karpeles, Joann Kealiinohomoku, Juana de Laban, Portia Mansfield, Samuel Martí, David P. McAllester, George P. Murdock, I. I. Potekhin, Curt Sachs, Ted Shawn, Estelle Titiev, Frank Turley, K. P. Wachsmann, and Richard Wolfram. Wolfram's contributions were especially substantial.

The first version of this survey, submitted on September 30, 1958, took the form of a symposium among correspondents. By July, 1959, at the suggestion of the Editor, it had taken the present form of an essay, which was then sent for additional CA☆ treatment to eight scholars of whom the following returned comments: Erna Gunther, Fred Eggan, James H. Howard, T.F.S. McFeat, Ted Shawn, and Richard Wolfram. Only contributions of these final commentators are identified by a star. The author and editors wish to express their thanks to the many correspondents and commentators who collaborated on this manuscript.

The exchange of books of Hungarian folk dances containing motifs and whole dances written in Kinetography, against copies of my book and the scores of dances published by my Institute in Essen, is continuing. . . . We have received orders from both Eastern and Western Germany, Yugoslavia, Hungary, Holland, Sweden [for] scores of national and historic dances [1956a].

In Yugoslavia, the Laban system was introduced by Prof. Pino Mlakar . . . the Yugoslav folklore institute has accepted Kinetography as the official method of notation. The first collection of Yugoslav dance scores written in the Laban system has been published.

In Hungary, due to the energetic work of Emma Lugossy and Maria Szentpal . . . the Corpus Musicae Popularis Hungaricae contains a large section in Kinetography, and several folk dance collections containing scores have recently been published. In Poland, Prof. Stanislaw Glowacki advanced the cause of Kinetography in the thirties, and recently a Kinetography group has been formed under the leadership of folklorist Roderyck Lange of Thorn . . . teachers of Kinetography have been trained at the Kinetographisches Institut, Diana Baddeley from Great Britain, Helmut Kluge from West Germany, and Ingeborg Baier from East Germany [1958a].

According to Juana de Laban (letter, Sept. 19, 1958), the Surrey Laban Art of Movement Centre has published national dances of Yugoslavia, Israel, and Austria.[2]

In Asia it is the dances of India which have received most study, especially the art forms, although lately the folk dances have been described as well. In other parts of Asia, save in Bali, theater dances rather than remote folk forms have drawn attention. Dances of Oceania need systematic study, except for the hula. Australian aboriginal dance has been reported only in ethnographic studies; a team of trained dancers produced a ballet and travel book about Australia instead of a much-needed analysis (Dean and Carell 1955).

The spectacular dances of Africa have been studied piecemeal, in connection with research on music. K. P. Wachsmann, a leading musicologist, is optimistic that his new role as Scientific Officer of Anthropology at the Wellcome House, London, will lead to an integration of music and dance study, at any rate in Uganda (letter, Oct. 20, 1958).

In the Americas, most publications use verbal description only. But there are centers of Labanotation in Cuba, Brazil, Argentina, and Chile (Solari 1958), and in New York, Philadelphia, Boston, and other cities of the United States. Andrew Pearse is trying to elevate the prevailing approach toward Caribbean dance from sensational journalism to serious folklore study. In South American countries, folklorists are feverishly collecting and interpreting not only popular dances but also acculturated and indigenous rites expressed in dance. Especially in Venezuela, Brazil, Argentina, and Bolivia, they have been aided both in research and in publication by government agencies. Remote tribes, however, are generally left to ethnologists, missionaries, and adventurers. In Mexico, research has progressed spasmodically, depending on Government attitudes, since a boom in 1922. A recent bibliography of Latin American dances includes a detailed account of research activities and sponsors (Lekis 1958).

The bulk of the dance publications in the United States deal with European folk dances or European derivatives such as squares and longways, and some favorites appear repeatedly. But Latin American dances are now popular in dancing schools, and jazz dance, until recently relegated to collections of ballroom dances, is rising to the status of subject for serious research. Dances of the American Indian have been included in a number of ethnographic accounts, though, unfortunately, choreographers were not engaged by the dance-enthusiast Speck, or by the great teams that studied the Plains Sun Dance and Societies, under Clark Wissler and Robert H. Lowie. Further, Mason (1944) and others initiated a great vogue for distorted Indian dances among countless groups of Boy Scouts and interpretive dancers.

In Canada, the disparate British, French, and Indian traditions have been discussed together in popular lectures by Barbeau. Also, there is a miniature manual of French quadrilles (Lambert n.d.) and a book of children's rounds (Barbeau et al. 1958). Other anthropologists have investigated some of the native tribes in the vast interior expanses of the country. Some repeatedly beat a trail through Six Nations Reserve, and a choreographer followed in their wake (e.g., Speck 1949; Kurath 1951, 1954, MSc, d). On the West Coast, William Holm is justifying Erna Gunther's encouragement in the reconstruction of Kwakiutl dances. Despite picturesque accounts by ethnographers, choreographers have not ventured into the Eskimo's bleak habitat. As the shamanistic ceremonies and mimetic festivals retreat before the white man's ways, chances for comparisons with the also unstudied dances of the Arctic fade, and chances for the analysis of Eskimo square dances improve.

OBJECTIVES OF DANCE ETHNOLOGY

Notwithstanding the energetic collecting of folk dances within the last fifty years, we have but now arrived at a point where we can begin to define for dance ethnology the subject matter, the scope, and the procedures of this emerging discipline. So far, the views of its devotees are characterized by diversity and much disagreement.

The first question that requires discussion is: What is the subject matter of dance ethnology?

Ethnology deals with a great variety of kinetic activities, many of them expressive, rhythmical, and esthetically pleasing. Would choreology, the study of dance, include all types of motor behavior or only restricted categories? If the latter, what identifies "dance," which uses the same physical equipment and follows the same laws of weight, balance, and dynamics as do walking, working, playing, emotional expression, or communication? The border line has not been precisely drawn. Out of ordinary motor activities dance selects, heightens or subdues, juggles gestures and steps to achieve a pattern, and does this with a purpose transcending utility. When walking attains a pattern, it becomes a processional, which is treated as dance by Wolfram (1951: 54–56), Kennedy (1949: 84–90), and others. A utilitarian activity like rice-planting, rhythmical and often accompanied by song, can easily be stylized into dance (Moerdowo 1957; Shawn 1929: 170). The transformation from occupation to mimetic dance has often been achieved, by processes ranging from imitation to abstraction (Kurath, in Leach [1949–50] 1: 277; Lawson 1953: 11; Wolfram 1954). For example, in the codification of gesture into Plains sign language (Tomkins 1929), gesture remained utilitarian and formally haphazard, while in the choreography of Pueblo Indian Tablita dances (Kurath, in Lange MS; Kurath 1957c), gesture was idealized and integrated into a structure of song, symbolic text, and group movement. In a strict sense, dance ethnology would be confined to patterned phenomena. In a broader sense, it could deal with any characteristic and expressive movement, since everyday motions are the roots of dance. Pursued according to the strict sense, this newest ethnic science would have limited usefulness. Pursued according to the broader sense, its findings would be indispensable to all holistic cultural analyses.

A second, closely related, question is: What is the scope of dance ethnology?

Existent definitions of the science contain varying emphases on ethnic and choreographic content. Franziska Boas calls dance ethnology "a study of culture and social forms as expressed through the medium of dance; or how dance functions within the cultural pattern" (letter, July 30, 1958). Ten years ago Kurath made "ethnochoreography" synonymous with "dance ethnology" and defined it as "the scientific study of ethnic dances in all their cultural significance, religious function or symbolism, or social place" (in Leach [1949–50] 1: 352). (See Gunther [2].☆)

Kurath's definition includes a controversial term, "ethnic dance." In her letter, Franziska Boas identifies ethnic dance with "folk dance." Another dance educator's definition of folk dance would apply equally to ethnic dance: "Folk dance may be defined as the traditional dances of a given country which have evolved naturally and spontaneously in conjunction with everyday activities and experiences of the people who developed them" (Duggan et al. 1948: 17). But Wolfram includes folk dance in folklore, "the lore of historical high cultures, to which we have direct access, as we belong to them" (letter, Aug. 22, 1958). A performer, La Meri, restricts the term "folk dance" to "communal dances executed for the pleasure of the executant" (in Chujoy 1949: 177). Chujoy defines folk dance as "dance created by a people without the influence of any one choreographer but built up to express the characteristic feelings of a people" (1949: 191). Thurston defines several categories of folk dances (1954b: 4–5):

(i) Dances of folk-lore. The narrowest use. They include religious and magical dances, occupational dances, war dances, and so on.
(ii) Dances of the folk. Includes (i) and also popular recreational dances, but not skilled step-dances.
(iii) Traditional dances. This will differ from (ii) in including step-dances.
(iv) All non-professional dances. The broadest use.

A class of dances which many people would exclude ... is the fashionable ballroom dance.... But now there is ground for a real difference of opinion.

The American dance-pioneer, Ted Shawn, makes "ethnic dance" subsume "folk dance" as a subspecies, and further distinguishes "ethnologic" or art dance: "I have included pure, authentic and traditional racial, national and folk dance as 'ethnic' and the theatrical handling of them as 'ethnologic' and the free creative use of these sources as raw material as 'ethnological,' but there is no hard and fast rule, and no clear dividing line" (letter, July 5, 1958). Franziska Boas would "make a distinction between professional dance as distinct from secular folk, much as you might between art and craft" (letter, July 30, 1958). La Meri defines ethnologic dance as "those indigenous dance-arts which have grown from popular or typical dance expressions," excluding folk dance (in Chujoy 1949: 177), but she admits that folk dance "is the dance from which inevitably grows both in technique and spirit the dance-art of a nation" (1948: 33). Thurston (1954b) completely excludes from "folk dance" the commercial dances of the stage and screen, on the ground that the mercenary objectives of commercial dance remove it several notches further from the roots than art dancing.

Ten years ago, Kurath (in Leach [1949–50] 1: 276) inclined to an identification of ethnic and folk dance similar to Thurston's second category above. At present, she would restrict "folk dance" to secular forms, no matter whether of ritual origin; include all types, both secular and ritual, under "ethnic dance"; and agree to the distinction between folk or ethnic dances and art creations.

This still does not settle the question of the scope of dance ethnology, nor have we meant to imply a restriction of scope to ethnic dance. Most students would agree with Boris Romanoff that "the art of ballet cannot be carried into the domain of ethnologic research" (in Chujoy 1949: 92), yet ballet developed from European court dances, just as present-day Japanese Nō drama has its roots in early religious and secular ceremonies (Kurath, in Leach [1949–50] 2: 794). Similarly, while jazz dance originated with the people, it has been adapted to the stage. Nevertheless, jazz, and also modern creative dance, express significant facets of our way of life that are not expressed in square dancing. A culturally complete picture should, as in Miss Boas' definition, include all of these.

This point of view has a champion in an eminent dance historian and musicologist who has considered all forms of dance in his research. Sachs says, "The question whether ethnology includes all forms of the dance must be emphatically answered in the affirmative" (letter, July 7, 1958).

A third question on which there is published disagreement is that of the extent of a need for dance ethnology within the broader field of general ethnology. In part, the answer to this question must await further inquiry into the function of dance in culture, which, in turn, depends on more findings on the relative significance of dance in particular cultures. Scholars have justified their studies on dance, not only by their use to readers in search of information or of material for performance, but also by the functional significance of dance in society.

Thus, Sachs points out that the dance has aided sustenance and well-being (1933: 2). Cherokee dances, like many others, are prophylactic and contain "the principles that insure individual health and social welfare" (Speck and Broom 1951: 19). The Samoan dance aids education and socialization, because it "offsets the rigorous subordination" of children and reduces "the threshold of shyness" (Mead 1949: 82–83). A statement by Mansfield about the Concheros holds for many other peoples as well: "The dance is the most satisfying expression of their religious feeling" (1952). Such motivations of utility or religious feeling are confirmed by many writers, among them Kirstein (1935: 2): "The subject matter of primitive, or source dances are the seasons of man's life, the seasons of vegetation, and the seasons of the tribe's development or mythic history." Two experts on Yugoslav folk dances hold that, "Folk dances . . . composed the dramatic element of various rituals and actions, each of which had for the man of a primitive society significance of a ritual magic action" (Jankovic and Jankovic 1934–51: 48).

In Western culture, religious content is increasingly relegated to art dance, while folk-dance activities are largely recreational (Mayo 1948: 3; Holden *et al.* 1956: v), or they are educational in that they break down ethnic prejudices (Herman 1957: v) and can, thereby, contribute to world unity (Shawn 1929: xii). This shift of purpose from faith to fun spreads inexorably to other parts of the world, as our dances spread; it widely changes public attitudes in the direction of exhibitionism. "First nights" succeed "first fruits" (Singer 1958: 379).

FIRST CIRCUIT

COMMON PROBLEMS OF CHOREOLOGY AND ANTHROPOLOGY

Choreology recognizes the cultural setting of dance, including the cultural position of individuals and of the sexes, and patterns of social organization and economic activity. It can identify local styles and styles spread over larger areas. Further, choreologists can design comparative studies to solve problems of prehistory, orthogenesis, diffusion, and internal and acculturation changes.

SOCIAL RELATIONS

Individual and Group: Creativity

The individual dancer's role within the group-dance pattern is a matter of local custom. In relatively few places, he can exercise his creative imagination uninhibited; more often he is submerged within the traditional group pattern, or permitted only some leeway.

In native America, limited freedom is reported by many observers. For example, ". . . every Conchero takes his turn in leading the dance. He can introduce new steps and if the Jefe approves of them they may become favorites" (Mansfield 1952: 151–52). Again, "The Iroquois Eagle Dance illustrates the pattern phenomenon in ritual and it permits the free expression of personality within set forms" (Fenton and Kurath 1953: 75). But contrasting practices between the Woodlands and Plains, and between the Pueblos and the Northwest Coast, are evident. In the Woodlands, "All of these circular and linear formations involve the cooperation of a group, commonly the whole community. . . . Across the Great Plains and westward, individual exhibitions are at least as popular as group formations" (Kurath 1953: 64). "The joint dances of the Pueblo Indians in which participate a large number of dancers dressed alike and in formation, are quite foreign to the North Pacific coast where the single dance prevails" (Franz Boas 1955: 346).

The role of the individual dancer may vary within a community, according to the function of a dance: "A sacred-profane dichotomy is still characteristic of the Pueblos. . . . [In secular dances,] there is no limit to improvisation and to the introduction of novel forms, whereas such innovations are strongly discouraged and controlled in other ceremonies" (Dozier MS: 149).

In modern America, the set patterning of square dances (Mayo 1948) contrasts with the almost chaotic freedom and formlessness of jitterbugging (Kurath and Chilkovsky 1959; Kealiinohomoku 1958).[3] (See Fig. 4.)

Elsewhere over the world, we find similar variety, with freedom generally a male prerogative, however, as in Yugoslavia: "Single folk dancers who are phenomenally gifted introduce into the collective style something of their own individuality. This must remain within the frame of the collective technique" (Jankovic and Jankovic 1934–51: 30). According to Hamza, the Ländler and Schuhplattler dances of Bavaria and Austria permit improvisation only to the male: in former times improvisation was imaginative; today the forms are more stereotyped (1957: 23). (See Fig. 9.) In Samoa, dance is still individualistic, being especially free for the boys: "It is a highly individual activity set in a social framework" (Mead 1949: 78). It is remarkable how male exuberance is similarly expressed in these otherwise contrasting cultures through stamping, leaping, clapping, and slapping of the thighs (Hamza 1957: 24–25; Mead 1949: 80–81).

Male-Female Roles

Let us look now at male-female relationships, first in terms of exclusive societies and the effects of exclusiveness on dance patterns, and then in terms of mingling of the sexes in dancing.

Anthropologists have reported the initiation rites of male dance societies. In addition, a theater expert has made a study of dramatic and choreographic patterns in the masked rituals of Patagonians, Australian Aborigines, and other primitive people (Eberle 1955). These rites, from which women were excluded or by which they were frightened, have now mostly died out. Analogous warrior ceremonies in the North American Plains are also in the throes of extinction or survive only in the War or "Grass" Dance. The reports for these rites are unfortunately too sketchy for reconstruction of their dances (Wissler 1916). In Europe, the traditional ceremonies of surviving male brotherhoods have been carefully described, and sometimes analyzed, for instance, the Calușari of Rumania (Wolfram 1934; Sachs 1933: 227–28), the sword dancers of Austria and England (Wolfram 1951: 66, 82; Kennedy 1949: 60–77), and the puberty rites of Hessen-Nassau; both in Europe and in the Plains, the societies emphasize age-grading (Zoder 1950: 87–90; Wissler 1916).

Women's ritual societies are absent in many cultures. However, in Yugoslavia the rain-bringing Dodole are young girls who dance from house to house (Jankovic and Jankovic 1952b: 13). Austrian female societies function at weddings (Zoder 1950: 91–92). The Iroquois have some women's dancing societies (Kurath MSa), as do Southern Plains tribes like the Ponca (Howard and Kurath 1959: 5).

Males generally dominate ceremonial activities. Of the ten Kiowa tribal dances, Gamble says that five are exclusively for men, and none are for women alone (1952: 100). Among the Iroquois, on the other hand, the women are more prominent. And in Brazil, Almeida voices surprise at the monopoly of females in the rite, Bumba-meu-Boi, both as chorus and dancers (1958a: 12). In Asia, their prominence varies in adjacent countries, and even within a single country. Thus, in the Cambodian dance dramas, girls play both princes and princesses (Shawn 1929: 160), but in Japan and China, female roles are usually played by men (Shawn 1929: 51). In Japan, again, it is priestesses who "perform the type of dancing called Kagura" (Shawn 1929: 37–38).

Customs governing the participation of the sexes recur strikingly throughout the world. To give just one example, for Yugoslavia and the Iroquois, specialists have made identical observations as to separation within a line, alternation, or pairing: "In earlier times men and women danced in separate kolos, but the female kolo was led by a man. . . . Later they danced the same kolo but grouped by sexes" (Jankovic and Jankovic 1934–51: 14). "Each sex fulfills specific ceremonial assignments and enters the dance in a prescribed order. . . . Social dances pair the sexes; most ritual dances segregate them in a line. . . ." (Kurath 1951: 124).

The arrangement of the sexes may provide a clue for relative chronology, for observations from various parts of the world seem to corroborate the statements from Yugoslavia and the Iroquois. Initiation and shamanistic rituals are danced by men alone or women alone (Eberle 1955: 186). Wolfram (1951: 82) considers such rites and their European survivals very ancient. In agricultural ceremonies men and women participate but without contact, as in the Pueblo Corn Dance (Kurath 1957c). But in social dances the mingling has become more intimate (Speck and Broom 1951: 66; Gamble 1952: 102).

Though courtship is one of the most fundamental of human activities, couple dances with courting mime appear to be fairly recent. In some couple dances, the man and woman hold hands, perhaps with many intricate arm figures, as in the Austrian Steirischer (Wolfram 1951: 180–83) or the Renaissance Allemande (Horst 1940: 31–40). The embrace dance evolved in Central Europe from the sixteenth-century Weller to the nineteenth-century Waltz (Sachs 1933: 184ff., 257). In American Indian social dances, the man may place his hand on the woman's neck or shoulder (Martí 1959: 143). In modern Pan-Indian couple dances, partners cross arms or lock elbows, but probably as a result of White influence (Kurath 1959: 34).

Organization

Ritual, but not secular, dances are generally organized according to the objectives of the enacting group, be it a small closed ritual society, a hereditary division, an economic group, or some combination of these.

1. *Ritual Societies.* The male societies discussed above have traditional systems of officeholding. Sometimes officers serve for life; sometimes they are elected periodically. The roles of the officials find choreographic expression, and are often identical with leading roles in professional dancing guilds (Wolfram 1934; 1951: 75–85).

In Mexico these organizations tend to be elaborate and often hierarchic, and the reflection of these features in choreographic grouping is evident to both ethnologist and dancer: "The Yaqui have a closely knit organization with three major divisions. The first is the church organization proper, which appears to be allied to the matachin dancers' society. . . . Next is the dancers' society, and thirdly is the fariseo (dancing clown) society" (Beals 1945: 107). "The Yaqui Matachini are led by a monarca and each of the files by a monarca segundo; the rank and file are called soldados" (Kurath 1952: 237).

The hierarchy of the Concheros society has a multiple function: "The organization is military as to discipline and titles, and a religious brotherhood as to purpose and the vows and obligations of members" (Mansfield 1952: 144). The captains also lead the dances (Kurath 1952: 237).

2. *Clan and Moiety Organizations.* Though perhaps modeled on civic organizations, and even linked with them by an overlap in leadership roles, ritual societies remain discrete. However, sometimes ritual organization follows hereditary and antithetical clan or moiety divisions. This phenomenon has survived among North American Indians such as the Creek, Yuchi, and Cherokee in the Southeast, the Iroquois and Musquaki in northern U. S., the Pueblos of the Southwest, and others, and among some of these peoples it finds choreographic expression.

For example, Speck observes of Iroquois dance forms that, "On each 'side' or moiety, there are two groups, the one of males, the other of females" (1949: 39). Similarly, the Seneca Eagle dancers, paired by opposing moieties, "always face their partner" (Fenton and Kurath 1953: 233).

Such relationships affect the spectacles and plaza circuits of the Pueblo Indians as well. In ethnographic terms, "Each Tewa pueblo is comprised of two divisions or moieties," which successively govern the ritual observances appropriate to the summer and winter halves of the ceremonial year (Dutton 1955: 6). In choreographic terms, "The Tewa moiety pattern has a profound effect on the circuits. . . . In San Juan the winter and summer moieties operate in harmony. . . . Santa Clara is split into four parties. All parties follow a set circuit" (Kurath 1958a: 24–25).

In other parts of the world, clans and not moieties are the significant hereditary divisions, as in Scotland, and clan totemism produces complex rituals, as in Australia (Eberle 1955: 427–53). However, the relationships between clan divisions and dance patterns have not been clarified, whether because of non-existence, or non-observance by field workers, we do not know.

3. *Economic Groups.* Economic specialization often creates occupationally disparate groups with special

dances, dance organizations, and dance functions. Gorer distinguishes these characteristics in African communities: "The best dancers come from the smaller, hunting tribes. In the larger, agricultural tribes dance diminishes in importance and vitality" (1944: 34). Tax observes that a Guatemalan community "that specializes in maize may pay more attention to the ritual aspects of culture—rain, planting, and harvest ceremonies.... Industrial communities should show special characteristics in contrast with agricultural communities.... They should be more secular-minded since they are less dependent on the vagaries of nature and more dependent on trade and money" (1952: 63). Agarkar emphasizes the superior dancing of lower castes in the Maharashtra of India, and the difference in dance types among "three main groups that are culturally distinct, namely, the Brahmins and the other advanced classes, the agriculturalists and allied tribes and the hill-tribes" (1950: 3). In Austria, craftsmen and miners retain occupational dances.

Urban-rural and class differences in dance forms exist in the West (Kurath 1952: 237; Lekis 1958: 90), but they are not so sharp as in India with its caste system. Thus in Brazil and Cuba, city has borrowed from country, with mutations (Lekis 1958: 202, 238), and in Europe, the court took from the peasantry (Horst 1940: 99–105), or the two engaged in give-and-take (Kennedy 1949: 102).

CHOREOGRAPHIC AREAS

A survey of dances over the world—or over even a limited area of the world—reveals dazzling variety, as suggested by the following comments on, respectively, Africa, Madagascar, and Greece. "African dancing ... involves the whole body ...; the dance steps are above all acrobatic" (Gorer 1944: 19–20). "The movements ... are very largely of the hands with rather minute sorts of gestures with the fingers" (Danielli, in Thompson 1953: 41). "The predominant feature of the dances is the variation of rhythm in slow and quick steps" (Crosfield 1948: 23). (See Fig. 5.)

Most students of continental forms have preferred to treat their subject according to political units, e.g., by countries of Europe (Alford 1948–52) or by states of Spain (La Meri 1948). Still other choreologists have delimited studies in terms of linguistic groupings (Lawson 1953) or topographical surroundings (Agarkar 1950: 3). In separate publications, Kurath, Mason, and Mansfield have agreed on the general characteristics of Amerindian dance as distinguished from European-derived dance of the New World; but Mason considers inter-tribal differences inconsequential (1944: 9–10), while Kurath is impressed by the cultural implications of these differences and by the usefulness of area mapping.

Fortún de Ponce (1957) shows Bolivia's regional differences on a map. Dances of native North America may be simpler to map than those of other parts of the world, where castes or racial mixtures complicate the picture, or where surveys are to date still sketchy. However, such mapping is highly desirable and could be carried out in terms of a number of criteria, the most fruitful being that of ecology.

It is certain that environment affects the repertoire, the content, and the form or style of ritual and dance. Common or similar natural resources have produced similar beliefs and dance rituals, for instance, in the Great Lakes region (Kurath 1957a: 1). Primary dependence on a certain commodity, such as maize or buffalo, everywhere inspires dances appropriate to that commodity. Subsequent tribal shift to a new environment enriches the repertoire, witness the Musquaki (Kurath MSd) and the Plains-Ojibwa (Howard MS: 24), both of whom migrated from the Woodlands onto the Plains and took on buffalo ritualism while retaining a residue of Woodland dances.

Such dances may well persevere after the extinction of the object-species, as does the Iroquois Passenger Pigeon Dance at the modern Spring Maple Festival (Fenton 1955: 1–2), or after a new religion has rendered the rites meaningless (Slotkin 1957: 13). On the other hand, the repertoire might change with the loss of aboriginal occupations such as hunting (Gamble 1952; Kurath 1958b). Studies focused on this problem might reveal similar phenomena in parts of the world other than America.

However, an analysis of dance form requires more insight than does an inventory of dance repertoires. Encouragingly, formal and evaluated analyses exist for several ecological zones of the world. Agarkar (1950) has made a noteworthy contribution in relating ecology and dance form in Maharashtra. Lawson has suggested connections between topography and dance movement, such as the traveling movements of steppe dwellers and the leaps of mountain people (1953: 32). Wolfram has made similar observations about Austrian mountaineers and plainsmen (1951: 200). However, Sachs cautions against hasty generalizations about geographical differences, saying "scholarly research into this question is still remote" (1953: 53).

Explanations for these differences are even more hazardous. Ecology certainly is an important factor in determining occupation, ritual, and dance; and it may prove to have an influence on style. Kurath has observed similar dance formations among widely removed agricultural groups (1956a), but doubts the applicability of ecological criteria to finer points of style. In addition, other factors must be weighed. Racial characteristics suggest themselves as a criterion when one compares Gorer's comments on Africa with Courlander's remarks on the "gross, direct movements" of Haitians (1944: 35) and with Dmitri's statement about the "use of the whole body" in Brazilian dances (1958: 9). Such a criterion receives further support from Kealiinohomoku's comparison of African and American Negro styles (1958). Again, Lawson (1953: 202) suggests linguistic affiliation as a reason for choreographic similarities. Several writers have speculated on psychological causes: Holt and Bateson speak of searching for the "cultural temperament" that motivates leaps or shuffles (1944: 52); Sachs has voiced daring theories on the relationship among manner, temperament, and environment (1933: 128–29); and the Jankovics seek the source of style in the "spirit and character of the people" as well as in externals of "social, economic and political conditions, geographical, topographical, and climatic circumstances" (1934–51: 29).

RECURRENT FORMS: ORTHOGENESIS, MONOGENESIS, OR DERIVATION?

As impressive as local variation, and more baffling, is universally recurrent manifestation, e.g., of "Rituals for increasing food supply, augmenting raw materials, controlling the weather, and warding off natural catastrophes" (Titiev 1955: 404–406), and of corresponding dance practices (Kirstein 1935: 1). Three main theories have been advanced to account for this: (1) orthogenesis following parallel invention; (2) common archaic origin, and subsequent migration and local adaptation; (3) derivative diffusion by direct culture contact.

In given cases, definitive explanations are usually wanting, but they can be arrived at by several modes of reasoning. Obvious and simple, or superficial, similarities of form suggest (1), while agreement of many elements and of complex, unusual patterns suggests (2). Further, at times archaeological or historical facts afford helpful evidence. Thus, for the maskers and rounds of Middle America and Central Europe, Kurath suggests independent origin under analogous circumstances (1956a: 296). Again, Spence and Sachs have both historical and formal evidence for direct, ancient connection of the present "Pyrrhic dances of the Balkans, Southern Russia, and even Southern France, which was powerfully affected by Greek culture" (Spence 1947: 3). On the basis of element count, Sachs proposes Rumania as the center of origin for these pyrrhics (1933: 228). On the strength of an equally striking list of common elements, Kurath bases arguments for (3), that is, for the derivation of New World correspondents out of this Old World base. For more recent events, historical documents can substantiate theories of derivation (Fenton and Kurath 1953; Kurath 1956a: 287, 292). Modern folk dances clearly betray their history of derivation, through name, form, and documentation (e. g., Tolman and Page 1937).

DYNAMIC PROCESSES

Continuity

Thanks to a streak of conservatism, many peoples have retained dances unchanged through centuries, e.g., the Japanese (Matida 1938), the Yugoslavs with their ritual mime (Jankovic and Jankovic 1957: 53–57), and the Pueblo Indians (Dutton 1955: 6–16). All dances have roots in some ancient form, but some cling to the roots with remarkable tenacity. One of the most famous examples of the latter is the European chain-and-song-dance, of which Wolfram says, "A Greek picture from Ruvo (400 B.C.) shows a chain-dance at a funeral. In modern Greece as well as in southern Italy such dance-forms, looking exactly like the old picture, are still current and are called 'Tratta,' for instance, at Megara" (1956: 33–34). These dances have been accurately notated and rhythmically analyzed in Lattimore (1957) and Crosfield (1948: 8). (See Fig. 5.)

Diffusion

One dance has turned up "all over Europe, north, south, east, and west" (Alford and Gallop 1953: 51). "In England it is coupled with the Maypole" (Kennedy 1949: 97). In the district of Parnassus its steps are "the same as on the Faroe Islands" (Wolfram 1951: 90). Wolfram, like others, infers from the steps a relationship between the Faroe rounds, the Yugoslav kolo, the hora of Rumania, and the French branle of today and of the sixteenth century. Yet each of these dances shows peculiarities of local style. (See Fig. 10.)

Some dance phenomena have spread even farther—the hobbyhorse almost around the world during prehistoric times (Spence 1947: 143–44, 167ff.), the modern fox trot under our very eyes. Diffusion can be supported by documentary as well as formal evidence, as in the cases of the Eagle-Calumet complex (Fenton and Kurath 1953) and the Morisca type from the Old World to the New World (Kurath 1949; Lekis 1958). (See Fig. 11.) Examples of postulated diffusion are legion, though not all are equally tenable.

Many diffused dances have entered repertoires alongside forms persevering through continuity. If their origin can be ascertained, the amount of adaptation to local style can serve as a measure of the antiquity of this entrance. The fact that Pueblo Buffalo dances have undergone "Pueblo-ization" implies lengthy presence (Kurath 1958b). The American Pan-Indian dance complex betrays recent origin by its faithful adherence to one style, whether performed in Oklahoma or hundreds of miles away (Howard 1955).

Transculturation

Often borrowings are reciprocal, especially among adjacent groups, to the enrichment of the repertoire of each. One example from the Pueblos will suffice: "Tanoans gave the moiety concept, perhaps animal and hunt societies to Keresans while they received in turn medicine societies, katchina cult, perhaps the clown societies and some notions of the clan.... Plains-Pueblo borrowing is a lively process" (Dozier MS: 156). (See also Fig. 12.)

Acculturation

Frequently borrowing is largely one-way, and takes place more or less under compulsion. A potent influence in this process was the expansion of European political control. In the New World, spectacular effects developed as blends with, or adapted borrowings of, European forms. Kurath has discussed such ritual and secular resultants in Mexico (1952); Lekis has identified not only Indian-Spanish but also Indian-Negro-Spanish mixtures in many countries of Latin America (1958). Almeida says of Brazil, "Our popular folk ballets are of three origins: Portuguese, African and Indigenous. ... The blending was complete and these dramatic dances have been entirely remolded" (1958b: 145). (See Fig. 7.)

Kurath considers it possible to unravel the blends into their components. Mansfield voices some scepticism as to the possibility of identifying native and European steps in, say, the dance of the Concheros (1952: 252).

Enrichment

Change can take place spontaneously, through internal development, or as a result of contact with external forces. Huge and varied repertoires have accumulated through millenia of borrowings, as in Yugoslavia with its many historico-cultural layers (Jankovic and Jankovic 1934–51; Kurath 1956a). The Japanese Nō theater combined many cultural strains (Shawn 1929: 22–37), and, as is usually the case, these borrowings were voluntary. On the other hand, the Mexican Indians accepted, remolded, or created new forms under forceably imposed European influence (Santa Ana 1940: 128), while they abandoned many indigenous ceremonies. In the same way, forced migration, as well as conquest, can develop rich blends, but always with some loss, for instance, the rites of the Haitian Negro (Courlander 1944). The difference between these two processes furnishes a subject for future study in dance adjustments.

Decline

Perhaps all dances are destined ultimately to decline. Currently in all countries, conservatives lament the deterioration of dances under pressures of modern industrial civilization, e. g., Agarkar, Wolfram, Vega. The decline takes various forms: change in overt features, such as paraphernalia (Lange 1957: 72–73); change of function, such as from hunt to weather control (Kurath 1958b: 439); secularization (Gamble 1952: 95); "discrepancy between the ideal and the actual" (Slotkin 1957: 15); deterioration of performance quality (Kurath and Ettawageshik 1955: 3); and simplification (Sturtevant 1954: 64). This pattern, noted among American Indians, is paralleled among the Yugoslavs, according to the observations of the Jankovics (see "Second Circuit," "Hypotheses"), and may well apply over the world.

Resurgence

The phenomenon of resurgence appears in various guises. For instance, renewed enthusiasm for tradition has in Santa Clara and Isleta Pueblos produced "a resurgence of dancing . . . and old dances which have not been performed for years have been revived" (Dozier MS: 148). But in Israel, where immigrants from diverse countries are reviving the ancient Hebrew culture, splendid new folk dances have arisen "from Chassidic and Yemenite traditions, from the energetic Horas of the Balkans, from the Arab 'Debka' . . . a synthesis between Orient and Occident" (Kadman 1956: 166). In Austria, revivals have grown not only out of scholarly, conscious efforts, but also among the folk, for example, the Salzburg "Jakobischützentanz."

Rebound

Rare and entertaining is the rebound of a dance from a land of acculturation to its home soil. Kealiinohomoku, Chilkovsky, and Kurath can corroborate the statement by Herskovits that "the dance itself has in characteristic form carried over into the New World to a greater degree than any other trait of African culture" (1941: 76). Now, reports Jones, the young African neglects his own heritage for a Euro-African song-dance, Makwaya, and for adaptations of modern American ballroom dances (1953: 36–37).

Review

Interpretation of dynamic processes relies on a core of style and structure that is manifested not only in the dance patterns of a culture but also in the human relations, social organization, adaptation to environment, and adjustment to contacts of that culture. The unraveling of dance mixtures depends on separate analysis of both the intruding and the native patterns as well as on analysis of the mixtures. In this way do anthropological and choreographic aims and methods parallel one another.

RELATION TO OTHER FIELDS OF RESEARCH

Psychology

"The anthropologist can find in the study of the dance corroborative materials for his observations, as well as clues which will direct his research toward new aspects" (Holt and Bateson 1944: 52). This applies to normal behavior, to buffoonery (Beals 1945: 102, 129–31), to possession by deities (Sachs 1933: 35, 43; Herskovits (1950: 881–82), and to "holy dancing" of the American Negro (Kealiinohomoku 1958: 105).

From analysis of rhythm and tempo in curative Amerindian rites, it appears that the "effects are the very opposite from the derangement" (Tula 1952: 118–19). Again, symbols have been devised to denote mental states, for example, "extravert" and "intravert," and their corresponding dance postures (Loring and Canna 1956: 7–9). In combination, techniques and findings of this kind have unexplored possibilities for psychiatric treatment, as has been pointed out by both a physician and a dancer: "Dance as a therapeutic agency . . . is quite unparalleled in potentialities" (Lawton, in Chujoy 1949: 144); "The possibilities of dance as mental therapy must be explored" (Franziska Boas 1944: 6). Blanche Evan is training her pupils for treatment of neurotic patients and for preventive therapy (brochure, 1959).

Technology

Choreologic findings apply also to technology, especially to the study of work movements. Occupational dances can be magico-mimetic, such as the planting mime of the Portuguese Bugios (Alford and Gallop 1935: 117), or recreational, such as the Philippine Balitao or rice-planting dance (Shawn 1929: 170). They can fall differentially along a scale of stylization, from the realism characteristic of the Balitao to the abstraction that marks the mining symbolism in the Dürrnberger Schwerttanz (Wolfram 1954: 1–2). We have already remarked that occupational dances are numerous (Kurath [1949–50]: 277).

Anthropologists have studied traditional movements in work and in craft production from a practical point of view (Weltfish 1946), and have evaluated some components of the movements with regard to an economic transition to industrial techniques (Salz 1955: 111, 228). Lately, engineers have clamored for scientific roads to industrial efficiency (Gomberg 1946). Techniques for improving kinetic patterns are available and have had incipient application. Specifically, Laban's system of Effort symbols can be used to notate actions and to expedite efficiency training. The system depends on criteria of "Weight, Space, Time and Control of the Flow of Movement" for well-regulated, less fatiguing actions (Laban and Lawrence 1947: 406).

After all, motions of work and sport have the same dynamic components as those of the dance—"Swing, suspense, sustained resistance, percussive impact, thrust and throw, relaxation" (Prokosch 1938: 294).

Linguistics

There are two promising approaches to the connection between dance and the spoken word. One is the study of the "manifestation of the relation between language and symbolic movements, standardized in each cultural area" (Franz Boas 1955: 346–47). Among the recreational dance-mimes to words, a delightful example is Barbeau's collection of French Canadian children's rounds (1958: 17–18, 98, *passim*). Again, it is often possible to define the connection between the ideology and the patterns of gesture code, as with the Cochiti Tablita-dance gestures (Kurath, in Lange MS).

A second topic for research arises from the question: To what extent are choreographic similarities linked to original linguistic relationships or to later culture contact? Lawson believes that stylistic likenesses between the Finns and Magyars may be due to linguistic affinity, and the differences to separation and different environments (1953: 202–204). Kurath would attribute likenesses between the Iroquois and Cherokee both to linguistic and to cultural connections. She has observed that the long-separated Shoshoneans, the Hopi and Comanche, have different dances, but the unrelated, adjacent, Tewa and Keresan Pueblos share a similar dance culture. Evidently both factors are operative.

Mythology

The spoken word, mythological beliefs, and dance drama are often integrated. In Hawaii, the dancer chants his own accompaniment—"their great dances, the great hulas: they are from the gods" (Campbell 1946; 32)—and with symbolic gestures enacts the words. On the Faroe Islands, dancers enact tales of the Sigurd legend as they sing them. In Malabar and other parts of Asia, a separate chorus sings for the actors of the Ramāyana and Mahabharata (Bowers 1953: 64–87), much as the Greek chorus voiced dramatic words. In America, myths are rarely enacted, but a close relationship between imagery, song, "birdlike dance and the myths" is characteristic of Kwakiutl dances, for instance the *MatEM* (Gunther MS). The Iroquois False-Faces impersonate "the great fellow who lived on the rim of the earth, and secondly, his underlings, the common forest people," by kicking and sparring (Fenton 1941b: 401–402, 420). The more usual situation confronting the dance ethnologist in America is a mere explanation by Indians of a dance's origin, as in the case of the Iroquois *Ohgiwe*, "Death Feast" (see Barbeau 1957, side 2).

Theater Science

In a noteworthy contribution to *Theaterwissenschaft*, Eberle points out that primitive drama mirrors the people's *Weltanschauung*, their beliefs and mythology (1955: 538), and that it does so by a combination of dance enactment, dialogue, music mask, costume, setting, and lighting (pp. 18–19). He shows insight and imagination in his compilation of anthropologists' field materials and in his reinterpretation of them in terms of dramatic structure, though he relies entirely on verbal descriptions without choreographic symbols or analyses. He confines this work to Australian, Patagonian, and African primitive drama. Others have similarly interpreted the drama of other areas—Spence (1947) largely from literary sources, Alford and Gollop (1935) mostly from observation. While studies of historic medieval dance drama are many, structural analyses of Amerindian forms are few. Speck's study of personally observed Cherokee dance drama (Speck and Broom 1951) suffers from the same omission as Eberle's work—absence of dance and music notation.

Dramatic constituents other than dance and music are often included in choreographic textbooks, e.g., costume and paraphernalia in Alford (1948–52), Evans and Evans (1931), Jankovic and Jankovic (1934–51), Lawson (1953), and Sedillo-B. (1935). The natural setting or native architecture is related to the dances in Bowers (1953), Matida (1938), Dean and Carell (1955), Slotkin (1957), Wilder (1940), and Mansfield (1952). There are also specialized studies: many on masks (Fenton 1941b; see Bibliography to Kurath [1949–50] 2: 687); some on costuming (see notes in Kurath 1958b: 44); and one a penetrating analysis of Melanesian settings (Schmitz 1955). Gunther [3]☆ comments on these aspects; see also Figure 1 for integration of several aspects. Instruments, often described, are sometimes also related to the dance (Wilder 1940).

Archaeology and Art History

Students of early historic or prehistoric dance drama have consulted sculpture, painting, or architectural remains from those periods. Schmitz lists a respectable number of archaeological sources for his study of the development of ritual settings (1955). Hickman has reinforced his inferences from ancient Egyptian bas-reliefs by observing the contemporary dances of the same region (1957). Nellie and Gloria Campobello have found confirmation on Maya antiquities in the modern Maya movement style (1940: 13–27). Sachs, as reinforcement for his text, reproduces art works of prehistory (1933: Pl. 1), antiquity (Pls. 8–11), and the Middle Ages (Pls. 17–24). Europeanists have a secure basis for historic reconstruction of the Middle Ages and Renaissance in a combination of pictures, verbal descriptions, codes and notation schemes dating from those centuries, as well as modern survivals. Among those who have availed themselves of this rich heritage are Sachs, Wolfram, Gallop, Kennedy, Walter Wiora, and Maurice Louis.

Kurath believes in conferring with the archaeologists themselves, and approached Homer Thomas for chronology in connection with Eurasian ritual drama ([1949–50] 2: 946–47), Joffre Coe for Tutelo mortuary customs (1954: 161), and Bertha Dutton for leads on Pueblo dance prehistory (1958b: 447). Such co-operation could have two-way benefits if choreographers aided archaeologists to reconstruct ground plans and movement styles.

Musicology

While integration of choreology with most of the above-mentioned disciplines lags, the close association of dance and music are fully realized. Although there are dance studies without reference to music, and musicological analyses of dance music without kinetic references, a huge literature combines these arts. Textbooks for prospective dancers combine the steps with musical measures and counts, by means of various notation techniques discussed below under "Second Circuit." (See Fig. 8). Scholarly analyses go still further than such juxtapositions, particularly in the Balkans where prevalent rhythmic complexities have inspired mathematically exact concordances (Jankovic and Jankovic 1955). Kurath attempted a possibly too detailed analysis of the two arts and their synthesis in the Iroquois Eagle Dance (Fenton and Kurath 1953). Mansfield has demonstrated how "the Concheros steps fit the music so precisely that the playing of the concha becomes an integral part of the dance" (1952: 183).

Yet a vast field remains to be worked. For instance, no one has correlated the intricacies of jazz rhythms with motor responses. Also, no one has gone beyond mathematical counts in the attempt to base musical rhythms on the force and nature of the motor impulses guiding the dances, though Sachs envisions the following process: "Organized in regular patterns, motor impulses pass from the moving limbs to the accompanying music, only to revert from voices, clappers, and drums as a stronger stimulus to the legs and torsos of those who dance" (1953: 38). A very few authors have suggested relationships between emotions, expanse of movement, and melodic lines (Kurath 1954: 160; Lawson 1953: 52; Sachs 1953: 128ff.). In a noted musicologist's opinion, "In dance is found the overt physical expression in visual form of the physical and spiritual aspects of music" (Rhodes 1956: 4). This merges into esthetics.

Symbolics

Dance patterns, gestures, and paraphernalia transcend an ornamental purpose. Indeed, investigators have asked whether ideas symbolized in dance are also expressed by all other arts within a culture. Sachs and Wolfram have speculated on such a connection between arts and culture symbols. Sachs, with caution, suggests an inherent connection (1933: 116–17). Wolfram points out interlaced motifs in crafts and dance (1951: 34–35).

In a discussion of ground plans, Sachs recognizes a widespread, perhaps universal, symbolism in certain circuits, serpentines, and interweavings (1933: 99–119). Today a symbolic significance is attached to the formations of many ritual dances, even in Europe and America, but it has faded from recreational repertoires. How many participants in American square dances realize the original vernal meaning of their "Grand Right and Left" (Kurath [1949–50] 1: 290)? Experts in the *mūdras* of India can communicate, yet foreigners and even Indian government members fail to grasp this gesture code (Singer 1958: 369, 375). The ancient combat symbolism of Moriscas is not apparent to all spectators, and not even to all performers.

Such facts cast doubt on the "universal language" of dance. Yet choreologic methodology rests on the assumption of some universality, and must proceed on the basis of this assumption.

SECOND CIRCUIT:
CHOREOGRAPHIC PROCEDURES

Certain procedures used in choreographic analysis, such as observation, interviewing, consultation of secondary sources, and re-study, are shared with ethnologists. However, the analysis of dance requires additional, specialized devices. Dance constituents are only partially revealed by general verbal description. For comparative purposes, symbols are required. Such symbols have been systematized by dance specialists trained in kinesiology and art dance, but these symbol systems, with their respective types of analysis, apply equally well to ethnic dance and art dance, and also to utilitarian activities.

ANALYTIC RECORDING

The recording, and analysis, of dance movements entails breaking them down into their space and time components.

GROUND PLANS

Track drawings, with solid or dotted lines, are commonly employed to depict solo or group movement along the ground, though other schemes serve related but special purposes (e.g., Jankovic and Jankovic 1952a: 58; Hutchinson 1954: 84–87). For complex progressions by large groups, it is customary to draw a series of ground plans. For the identification of males and females, officials, dramatis personae, and the directions they face, choreographers have used diverse symbols, identified by a key (Duggan et al. 1948; Mason 1944; Vega 1952; Laban 1928; Hutchinson 1954; Kurath 1951, 1958a, b). Figure 1 shows an attempt to depict in one diagram the ground plan, setting, and dramatis personae of the Iroquois Eagle Dance (Fenton and Kurath 1953: Fig. 4 by Kurath). The dotted lines indicate the path of the Eagle Dancers. Labels A, B, A', and B' refer to the structure of the music and dance as related to the path of movement. The positions of the other actors are identified by numbers. Furnishings are spelled out. Ethnologists have often attempted similar diagrams for placement of participants or path of locomotion. Beals has successfully shown both placement and path (1945: 177–79).

Ground plans can be juxtaposed for comparison. Figure 2 illustrates a clockwise circle and a counterclockwise ellipse. In both cases the singer (S) is in the center. Such diagrams are instructive for area studies, e. g., for contrasting the paths of Great Plains and Woodland Indians' dances (Kurath 1953: 61–62). More complex diagrams can also be juxtaposed, but there is a limit to the amount of material that can be integrated on one page.

BODY MOTION

The graphic representation of three-dimensional body movement on a two-dimensional written or printed page poses problems. Textbooks for schools and the general public have usually bypassed this problem, and depended on technical terms in verbal descriptions, tabulating the verbal phrases with counts and beats. Their authors have often employed stick figures (Mason

1944; Evans and Evans 1931), full-figure drawings (Herman 1947; Mayo 1948), or photographs (Blanchard 1943; Agarkar 1950; And 1959), by way of graphic aids. Raphael Moya has drawn expressive stick figures for Campobello and Campobello (1940), and John Bancroft has given stick figures locomotor value in Lawson (1953). A device for showing footwork is a set of footprint outlines in numbered positions (Shomer 1943: 8, 14–15, 17; Vega 1952: 364–65, 369).

The inadequacy of these devices has for many years led to experiments with symbols. The most complete symbol system, and the one in widest use, is that of Rudolf Laban, termed "Labanotation" in the United States. It is not, however, the simplest one. Other ingenious systems have found useful application in field work and for special dance styles. None of these systems can here be explained in full, but the principles of Labanotation will be illustrated as applied to selected stylistic problems and comparisons. A few examples of other systems, which can combine dance notation with musical rhythms and structure, will also be given.

The Laban System of Notation and Stylistic Analysis

"In 1928, Rudolf Laban published a new system of notation in which he introduced the vertical, symmetrical staff, read from the bottom up and clearly picturing, for the reader facing the score, right and left, front and back. The other invention is . . . using the length of the symbol on the staff to indicate duration of movement" (Hutchinson 1954: 3; see Shawn [5]*). The level of movement—upward, downward or horizontal—is indicated by the shading of the symbol (Hutchinson 1954: 14; Knust et al. [1958] 2: 9–12). The Dance Notation Bureau of New York worked out some officially approved modifications and sponsored the publication of Labanotation for trained dancers (Hutchinson 1954) and for beginners (Chilkovsky 1955–56). Folklorists and anthropologists can find introductions to the system in Pollenz (1949) and in Juana de Laban (1954).

The Laban system records the motion of every part of the body. It can distinguish between a natural walk and various kinds of stylized walk. In Figure 3, (a) represents an ordinary, normal forward walk; (b) represents a shuffling walk, as in American square dancing and in some American Indian stomp or round dances; and (c) represents an elastic, syncopated progression, as in the Pueblo Tablita dances (Kurath 1957c). The system can also show creative or local variants. In Figure 4, individual, creative variation is illustrated by two versions of the Lindy jazz dance, from the notes of Nadia Chilkovsky. In Figure 5, local variation is illustrated by the Greek Syrtos and Kalamatianos, from the notes of Alice Lattimore. Variants of the notations in Figures 4 and 5, as well as symbols in Figures 3 and 7, were approved by the New York Dance Notation Bureau for publication in the *Dance Notation Record* ([1957] 8, No. 2: 6; [1960] 11, No. 1).

Style, or quality, can be shown more precisely by "diacritical" symbols for flexion or extension, accent, dynamics (strength), and effort. These useful symbols can be written separately as style indices (Fig. 6) or attached to a notation staff (Fig. 7). In Figure 6, the most basic style symbols (a) are compared with three other, more abbreviated, devices for showing posture variants.

Thus (a) exemplifies stick figures from Campobello and Campobello (1940: 21 [Maya], 89 ['Tarascan], 158 [Yaqui]), the left one indicating erect posture, the middle one indicating stooped posture, and the right one indicating greatly flexed posture. While the other symbols vertically aligned with these in each of the three cases indicates the same posture as the stick figures in the top row, (b) are symbols used in goniometry (cf. Kurath 1954: 160); (c) are symbols equated with extravert or intravert mental stages (Loring and Canna 1956: 9, key 1); and (d) are Labanotation extension and flexion symbols (Hutchinson 1954: 184, 264–65; Knust et al. 1958, 2: 5–6, 72–73, 78–79).

Figure 7 illustrates the compactness of the Laban system when applied to an acculturation problem in choreology. The illustration tabulates symbols for dance steps (co-ordinated with their rhythm), and for four aspects of style—posture in torso tilt and flexion, dynamics (degree of tension or relaxation), and effort (strong or weak impact). The problem is the discernment of native and Spanish elements in several hybrid Mexican Indian dances. Example (a) shows a typical step of the native base line in the Yaqui Indian masked Pascola dance (Kurath, field notes; Beals 1945: 120); (b) shows a step from the importation, the Spanish zapateado (Tsoukalas 1956: 19). The native qualities are forward tilt and slight flexion of the torso, relaxation, and fast, direct, strong effort; the alien qualities are erect and extended torso, tension, and fast, direct, light effort. The blends (b), a step of the Yaqui unmasked Pascola, and (c), a step of the Concheros, use the zapateado step, but maintain slight variants of the native qualities. The Concheros step receives a verbal description, with counts, in Mansfield (1952)—upbeat, slight lift on left foot, extended right foot; count 1, step on left foot to left side; count 2 brush right heel forward; count 3, step on right foot, closing to left foot. The efficiency in the use of symbols should be apparent.

Other Systems and Their Application to Musical Problems

All systems have attempted some combination of music with dance steps. The Laban system is convenient in this regard, for the dance symbols can be lined up alongside the musical notation. Several other systems have, however, achieved the graphic integration equally well, for example, those of Stephan Toth in Bratislava, Boris Zaneff in Sofia, Raina Katsarova-Kukudova in Sofia, Rudolf Benesch in London, and Eugene Loring in Los Angeles. The combination of dance notation and rhythm, specifically of long and short impulses, is shown in Figure 8, where (a) depicts a typical Pueblo Indian step (music fragment from Yellow Corn Dance of San Juan in *Midwest Folklore* [1958] 8, No. 3: 157), according to the Kurath system (see Kurath 1952, 1954; Thompson 1953: 35–38); (b) depicts the Mexican Jarabe Tapatio in the Sedillo system (1935: 14) and (c) depicts it in the Mooney system (1957: 28 [Fl means flat; b, ball; S, stamp]); (d) depicts the Bowarian "Zwiefacher"—the Ländler, with changing meter—according to Hoerburger (1956: 101 [D means "Dreher," or two-count turning step; W, waltz in three-counts]); and (e) depicts the

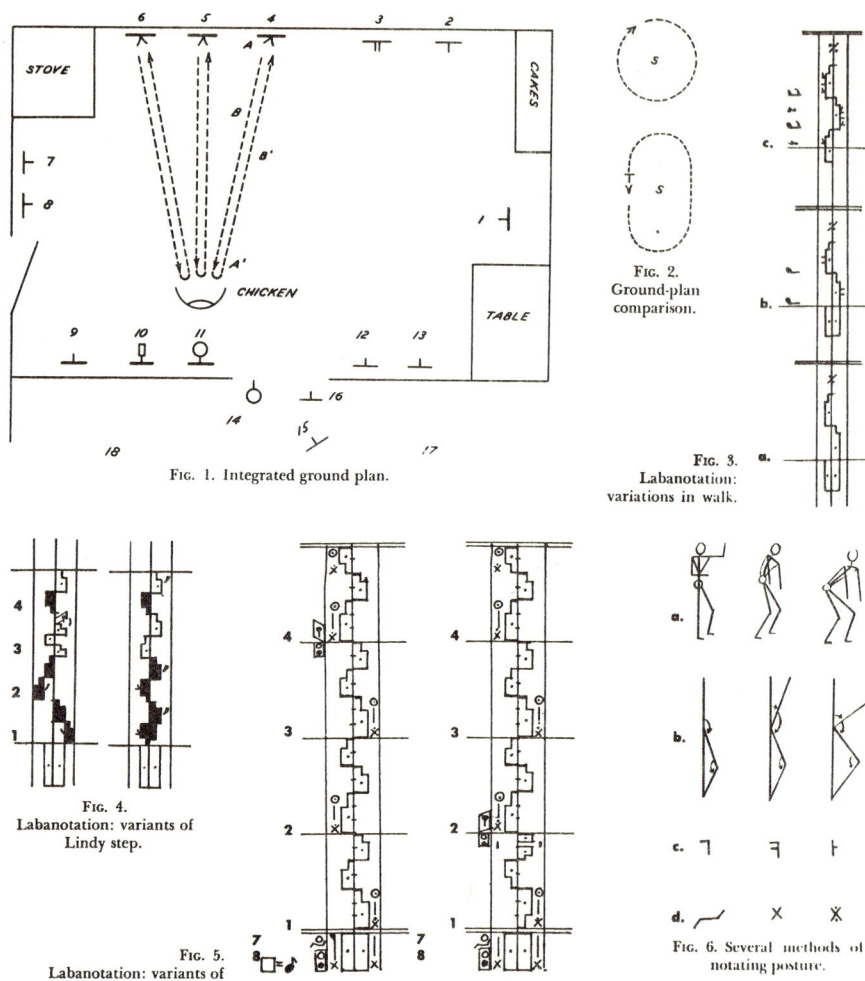

FIG. 1. Integrated ground plan.

FIG. 2. Ground-plan comparison.

FIG. 3. Labanotation: variations in walk.

FIG. 4. Labanotation: variants of Lindy step.

FIG. 5. Labanotation: variants of Greek Syrtos.

SYRTOS

KALAMATIANOS

FIG. 6. Several methods of notating posture.

Yugoslav Invertita, in the Proca system (taken from Balaci and Bucsan 1956: 224 [for principles, see Proca 1956, 1957]).

Several systems include devices for showing rhythmic stamps or hand claps. Ilka Peter and Herbert Lager devised one such method, for recording the stamps and leaps of the Perchten of Pinzgau, Austria. Figure 9 illustrates two methods for depicting hand strikes, on various parts of the legs or feet, in two male dances—the Austrian Schuhplattler (*a*) and the Rumanian Calušar (*b*). In Figure 9, (*a*) shows the "Sechserschlag" step in the Schuhplattler dance, according to the Horak system

(1948: 155), which places the dance rhythm on a musical staff (on counts 1 and 3, right hand strikes right thigh; on counts 2, 4, and 6, left hand strikes left thigh; on count 5, right hand strikes sole of left foot, while hopping on right foot). Directly underneath these symbols are the corresponding symbols (Greek letters, delta and sigma, for right and left) for these hand strikes, in the Proca system (1957: 87–92). These constitute (*b*).

The juxtaposition of dance notation with musical rhythm and/or melody can show the nature of phrasing. It can show, for instance, whether the dance and music phrases or meters coincide or overlap. Such relation-

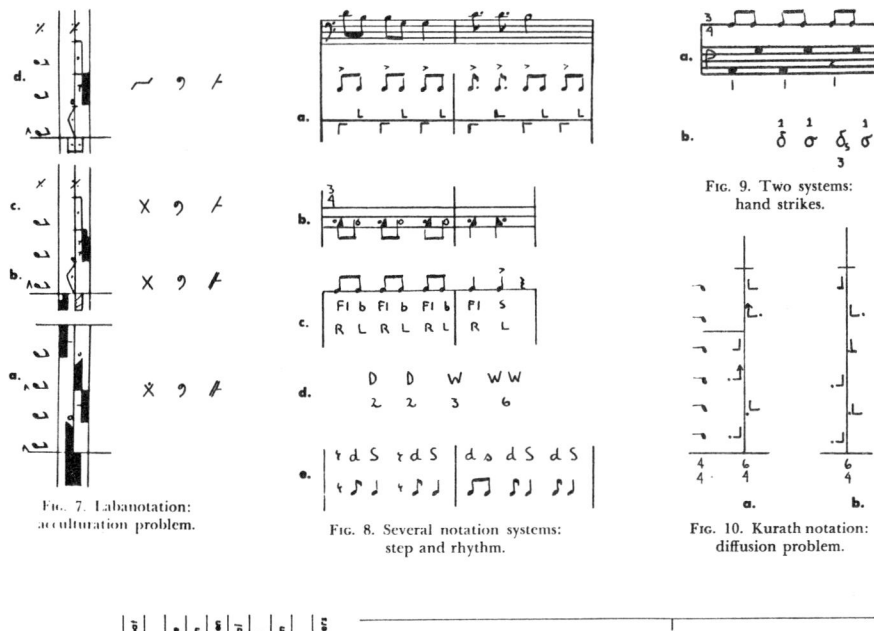

Fig. 7. Labanotation: acculturation problem.

Fig. 8. Several notation systems: step and rhythm.

Fig. 9. Two systems: hand strikes.

Fig. 10. Kurath notation: diffusion problem.

Fig. 11. Tabulation: symbolic elements in acculturation problem.

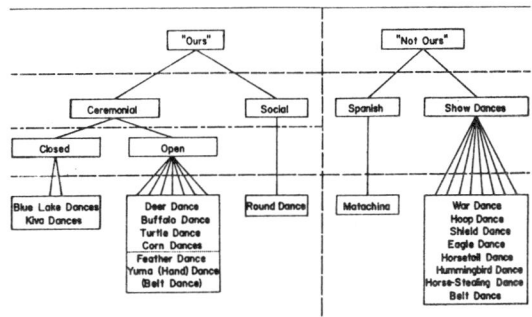

Fig. 12. Classification of Taos dances (1958).

ships are often important indices of local or national style, e.g., in German and Scandinavian dances, the meters usually coincide, while in the Balkans, Rumania, and Palestine, they often overlap. Once a local or national style is determined, by this criterion of phrasing, problems in diffusion can be investigated through comparison of the diagrams which reveal the nature of phrasing. Figure 10, for example, indicates diffusion of the Hora step to the Branle and Faroe step: (*a*) shows the phrasing of the Hora of Rumania or Palestine (step with left foot; step with right foot; hop on left foot; hop on right foot), while (*b*) shows the phrasing of the Faroe step and the French Branle to be essentially the same as that of the Hora, though a toe-touch replaces the hop.

The larger structure of a dance also can be analyzed in relation to the accompanying music, and with the aid of musical precepts, since, as a rule, the structure of the dance matches the music structure. Again, analysis of a folk dance can lean on methods for analysis and composition of art dance, since similar forms recur in both types of dance—the "binary" form of two themes (A B), the "ternary" (A B A), the "rondo" (A B A C A), and so on. After the basic form of the dance has been established, the dance score can be further labeled for its significant parts, such as phrases and larger sections. Art dancers have published explanations of these structural units, and their works can be consulted for details (e.g., Hayes 1955: 74–87). In dance ethnology, such structural analysis constitutes an interesting study for its own sake, and it can also illuminate problems of local style or of derivation and mixture of styles (Kurath 1954, 1956b, 1959; Kurath, in Lange MS).

Organization of Materials

Distribution maps of stylistic features of dance rarely contain notation symbols, despite the potentialities of symbols for this purpose. Thus, Kurath used other devices to tentatively map choreographic areas (1953: 69) and diffusion paths (Fenton and Kurath 1953: Fig. 30) in North America. Kennedy published a map of England's Ceremonial Dances (1949: 72–73). Several authors have provided pictorial maps of regional dances in Europe (La Meri 1948; Duggan *et al.* 1948: Alford and Gallop 1948–52; Wolfram MS; Zoder 1938: 174–75).

Tabulation has served as a clarifying device in many studies, though, like maps, without benefit of notation. Sachs made frequent use of tabulations for summarizing element distributions (e.g., 1933: 111). La Meri used tabulation in assigning Spanish dances to their various regions and origins (1948: 171–80), and Lekis in charting Latin American dances (1958). In an acculturation study of Mexican Moriscas, Kurath tabulated symbolic elements found in dances of Mexico, New Mexico, and Europe (see Fig. 11, condensed from Kurath 1949: 105–106). Kurath has also tabulated dance concepts and patterns ([1949–50] 1: 280–83, 288–93; 2: 748).

Classification of dances has been carried out according to a number of categories: region (Jankovic and Jankovic 1952b; La Meri 1948; Lawson 1953; Lekis 1958); period (Czarnowski 1950; Kealiinohomoku 1958); social class (Agarkar 1950); origin (Tolman and Page 1937; La Meri 1948; Lekis 1958); function (Kurath 1949; Speck and Broom 1951; Wolfram 1951; Brown 1959); form (Sachs 1933; Vega 1952; Holden *et al.* 1956); and ground plan (Howard and Kurath 1959: 8–11). Kurath's classification of Iroquois dances (MSa) follows Fenton's classification "into three groups according to their function of bringing man into rapport with particular spirit-forces" (1941a: 144). Brown (1959) has based his classification of Taos dances on native opinions (Fig. 12). Taos Indians have accepted Plains show dances, but have attitudinally kept them separate from their own religious dances. Within each category Brown distinguishes subdivisions, by sacredness or origin.

Evaluation of dance depends on both cultural and choreographic criteria, which vary from place to place. It hinges primarily on stylistic definition, but also on appraisal of factors such as creativity, precision, acrobatic skill, and dramatic ability. Slotkin (1955) devised a ruse for eliciting native standards with respect to a Menomini Contest Dance. Agarkar describes the dances of Maharashtra and publishes the dance-songs in Devangari music script "with a view to evaluating their social significance and to show the place occupied and purpose served by them in the complex social fabric" (1950: 12). Hein (1958) combines cultural and artistic criteria in evaluating the Rām Līlā of North India. None of these evaluations has relied on notation, either during investigation or for publication of findings.

Hypotheses

Reliable workers have generally hypothesized only in connection with specific problems; often the hypotheses have received subsequent confirmation. Thus Cecil Sharp's theories of the origin and function of the Morris dance have been confirmed and amplified by Alford, Gallop, Kennedy, Sachs, and Wolfram. At times, experts have evolved theories from regional observations which may have universal application. For example, the following findings of the Jankovics on change in ritual dances seem to pertain in parts of the world other than Yugoslavia: (1) change of dance rites, while choreographic elements are preserved; (2) change of choreographic and other elements, while ritualistic purposes are preserved; (3) change of ritual paraphernalia; (4) change of specified age and sex of the participants; (5) disconnection and dissipation of component parts of dance rites, which join and amalgamate with other parts and with folk dances (1957: 58).

Sachs has courageously speculated on symbolism of forms, on paths of diffusion, and other phenomena, prying into prehistory and history to correlate dance patterns with cultural layers, from earliest times up through the periods of Neolithic nobility and peasantry (1933: 148). Frances Wright has been working on a tabulation integrating dance, music, and religious forms through the ages.

Theories must rest on exact and manageable information. Solutions to questions posed above in the "First Circuit" regarding ecology, universality and diversity, psychology, artistic patterns, and practical application await systematic compilation of reliable data.

THIRD CIRCUIT: PRACTICAL CONSIDERATIONS

Past and future accomplishments depend, in part at least, on practical considerations such as the availability, cost, and differential usefulness of recording machinery; the training of workers in the field; and relationships between scholars and laymen.

Technological Aids

FILMS

Dance films cannot replace dance notation, any more than recordings can replace music notation. Even where finances permit photography, no film can supply the data for complete choreographies. Professional films usually cut both dance and music into fragments and obscure them with an educational drone; sometimes they are not authentic. Films by ethnographers often contain short dance excerpts which can serve as a reliable basis for some stylistic notation, but they are, after all, only excerpts. Films by choreographers, though perhaps technically inferior, are more likely to contain essential data because the choreographer tries to phrase the shots by dance sections and to co-ordinate them with notes taken on location. Although, in cases of unposed performances, they may miss some phrases during rewinds, Kaplan, Kurath, Kealiinohomoku, and Mansfield have made good use of such films for notations.

Techniques are being developed for viewing films at different speeds, repeatedly and backwards, and for halting at particular frames when necessary. Splicer viewers provide a simple tool for the scrutiny of single frames. Connecticut College School of Dance has a three-year Rockefeller grant to work out such techniques with a special projector (Rogers 1958: 5). The English Folk Dance and Song Society is experimenting with the use of films in stylistic analysis (Williams 1958: 111).

Problems relating to film techniques were discussed by the International Folk Music Council at its meeting in 1950, and by the Folklore Conference of 1950. Fragments of a conversation from the latter follow (Thompson 1953: 39–40):

THOMPSON: I was very much interested in Brazil to find that they were beginning a rather systematic collecting of dances ... of the State of São Paolo. And they were sending out sound trucks with movies. They were able at that time to make a record with sound and also with pictures of these dances, but they had not worked out the technique of co-ordinating these.

KURATH: I regret that we do not have more of these motion pictures and photographs of dances. There is a great technical difficulty in getting all around the dance to see from all points of view. There are two techniques.... In the old one you set up a camera and then take the dance from one angle. ... I think that is the best method for ethnologists. But nowadays there has been developed the artistic method of moving the camera around and getting shots from above and below ... we are mixed up with tabu.... You are not permitted to take movies of any of the ceremonies.

SEAN O'SUILLEABHAIN: We have done some recordings of certain types of folk dances in Ireland. We have the mummers ... we sent our gramophone to record them and also sent a camera two years ago where we made an hour-long film recording one of these dances. And we also took a record of the people speaking their parts. We had no difficulty whatsoever, and there is no tabu on recording the dance.

Still photography, monochrome or color, can have some value for the analysis of posture, costume and so on, especially if the pictures are taken in series. Such pictures are also more practicable for publication than are frames from films, and have been used to excellent effect in many books, for example, the beautiful set of Ainu dances in a book by Kawano (1956). They cannot convey the kinetic element, however, and thus can only supplement notes and films.

RECORDINGS

Field recordings on disc or tape have the same value for dance study as for music study, because they can supply the basis for the music notation that is vital to the dance score. Recordings are complete and can be used to fill structural gaps left in the films, though the reeling off of the film and the tape can co-ordinate only by chance. Kurath found that a simultaneous use of a recording and a film was very helpful in the analysis of Pueblo and Menomini dances.

Publication of interrelated records and books on dance has unfulfilled possibilities. Popular combinations for folk-dance groups have been sponsored by the Folk Dance House of New York and the Jewish Education Committee of New York, and great quantities of Folkcraft records have been produced to accompany a *Syllabus* by Herman and a pamphlet by Lapson (1954). The Folk Dance Federation of California has issued sets of interrelated discs and volumes of European and American folk dances. Ethnic Folkways records often are accompanied by dance descriptions. The Folk Dance House of Manchester, England, and other European centers have similar projects.

TRAINING OF DANCE ETHNOLOGISTS

"We should send trained dancers in company with mechanical technicians into all the countries where the native dancing is still true and untouched."

This vision of Ted Shawn's (1929: xiii) has not been realized for several reasons. First, suitably trained dancers do not grow on every bush, as do mechanical technicians. In fact, an adequate course of training—which should include courses not only in anthropology but also in kinesiology or "modern" dance (not ballet), folk dancing, dance notation, and music, preferably by the Dalcroze system of kinetic rhythmic analysis—does not exist. The usual graduate-study requirements in anthropology leave no room for three-year courses in choreography. Moreover, few schools offer the needed artistic combination. Those which do, include the Juilliard School of Music in New York and Rosalie Chladek's School in Vienna. Some have taught ethnic dance routines, such as Shawn's University of the Dance and La Meri's former Ethnologic Dance Institute. All colleges in the United States include modern dance in their physical education curriculum; many, such as Berea, include folk dance; some, for instance Swarthmore and Women's College of the University of North Carolina, have added courses in Labanotation. An increasing number confer higher degrees in dance. Some have accepted Ph. D. projects on ethnic dance, among them Texas Woman's University in Denton (Garcia 1958) and New York University (Mansfield 1952), where Sachs formerly gave a course in dance history. Nevertheless, colleges do not offer rounded courses or degrees in the combined disciplines. As a unique instance, John Mann offers a course on Ethnic Dance in the Social Science Department at Northern Illinois University. The course includes practical and theoretical training in history, forms, techniques, relation to music, and in Labanotation, with emphasis on dances of Polynesia and of the European Renaissance and Baroque periods, and with a more general survey of other styles.

Research that is sound must be done by dancers who have achieved the insight and point of view of the ethnologist, or by musicians and ethnologists with dance training. It is difficult for an anthropology graduate without previous dance training to learn dance theory and notation in a few spare hours. Even the simple Kurath shorthand cannot be learned in a hurry. Prospective field workers whose schedule cannot include choreography may resort to a check-list on dance in the *Guide for the Human Relations Area Files* (Murdock et al. 1954); to two questionnaires prepared by Kurath (*Midwest Folklore* [1952] 2: 53–55 and *American An-*

thropologist [1956] 58: 177–79); and to an amplified list being prepared by Kealiinohomoku.

SCHOLAR AND SCHOLAR

TEAMWORK

It would be a big order for one person to go to the field as choreographer, musician, photographer, technician, ethnologist, folklorist, linguist, and psychologist, and to continue as historian and archaeologist, not to speak of geographer and botanist. Consequently, teamwork is essential. Each member of a team may well master two of the above specialties, and they need not all enter the field at the same time, though this is preferable (see Gunther [4]☆ and Shawn [2]☆).

EXCHANGE OF IDEAS

Conferences

Mailed exchange is possible at a distance, but it is not conducive to alert repartee. Publication often lags years behind discovery, and may, even then, not become known or available to all who are interested. Personal meetings are preferable. The International Folk Music Council gives dance an equal place with music at its annual conferences in various countries. In the United States, the Society for Ethnomusicology and the regional and national folklore societies welcome dance topics. But American dancers rarely attend the former conferences because of the expense, and few belong to the societies. European scholars have profited from frequent meetings. Wolfram, together with forty other experts from Germany, Austria, and Switzerland, was active in the founding of a "Studienkreis Volkstanz" in 1957 in Stuttgart. Under its auspices folkdance leaders and groups from eight countries met for ten days in August, 1957, at Diest. Again, an international notation congress, held in Dresden during October 1–4, 1957, brought together scholars from many European countries. It was initiated by Wilhelm Fraenger, the Co-Director of the Institut für Deutsche Volkskunde an der Deutschen Akademie der Wissenschaft zu Berlin, and was organized by the Institut für Volkskunstforschung beim Zentralhaus für Volkskunst zu Leipzig. The frequent European folk-dance festivals also provide opportunities for discussion and seminars. In the United States, the National Folk Festival, managed by Sarah Gertrude Knott, has added small conferences occasionally. But, in the main, outside of Europe, dance scholars have had inadequate opportunities for interchange of knowledge. They have had even less opportunity for consultation with experts in the various anthropological sciences.

Centers

Another requirement for sharing knowledge is a clearinghouse for information. As in the case of ethnology, field materials on dance are strewn about in many archives and private collections, and manuscripts remain in private files unless some publisher has the courage to print them. Films, recordings, and even publications fail to reach interested students, though many items are available in the libraries of centers listed under "Selected Source Materials," below. To an extent, published items appear in various current bibliographies, in *Ethnomusicology*, and in the *Journal of American Folklore*. For the ethnological aspects of dance, the research scholar can refer to the Human Relations Area Files (headquarters: 421 Humphrey Street, New Haven, Conn., U.S.A.), which are available at several universities in the United States: full quotations on the dance are assembled mainly under Number 535, but also under related topics such as Gestures, Posture, Recreation, Theater, Therapy, Funerals, and Ceremonies. In 1959, a dozen American dance scholars formed the Dance Research Center, with headquarters in Ann Arbor, Michigan. The associates provide and receive information on research activities, publications, and films, and are planning several co-operative projects.

However, Latin American dance folklorists have received government support (see list of sponsors in Lekis 1958) covering both field work and publication. Some groups, such as the African Music Society, successfully turn to UNESCO. In the United States, individual scholars have some small chance at foundation sponsorship for research, but none for processing and publication. Yet the expense of the latter is high. For instance, the preparation of the score for a solo dance, in Labanotation, costs U.S. $40.00, and, for a group dance, U.S. $75.00, not to speak of clerical and other costs. But the Dance Notation Bureau is in a position—if financed— to aid individual field and publication enterprises, as Chilkovsky suggests (1958: 1):

We believe that the best assurance will come through a publications program broad enough to include . . . national and folk dance material, and composition studies. Cooperative interchange of ideas by the leading practicioners of movement notation will then result in a widely representative publishing outlook so that all notation needs on all levels and in all centers may be consulted. European countries have already released a wealth of notated national and folk dances. Why not publish books on American Indian dances, jazz steps, West Indian dances, Hawaiian dances?

In the interest of extending notation activity on a world scale, we should like to suggest that the Bureau or Music Publishers Holding Corp., or both, assume leadership in organizing an advisory council composed of notators in all centers, who can act as a clearinghouse for publishable materials. Such a council might be empowered to make recommendations to publishers on the basis of a broad overall perspective.

SCHOLAR AND PUBLIC

Another sponsor of publication must not be overlooked. That is the dancing public, which has determined the policy of many a publishing house and has prompted publication of works by Czarnowski, Herman, and other folk-dance specialists. The dancing public, which receives the information, may at times also provide recruits for research. Educators around the globe have found stimulus in lay demands.

The dance descriptions with their music, songs and social setting . . . are published at this time because of the great demand for this material during the three years of the California Centennial celebration, as well as to satisfy the eager desire on the part of California dancers to learn more about the pioneer dances of their state [Czarnowski 1950: 8].

It is due to this revival [in schools] that an attempt has been made in this book to put into written form some of the dance-lore, traditional steps and music of South India [Spreen 1949: vii].

Those of us who are engaged in making folklore available to the public are performing a fascinating, interesting and very valuable task, but one that has a good many difficulties. The greatest of these, I think, is that in many cases we are trying to take over a tradition that has developed unconsciously and graft it onto a conscious culture....

Now, as you know, the revival of folk music in England is mainly due to Cecil Sharp. He was not the first person to collect folk songs or perhaps even dances—but he was the first, or one of the first, to see the implications ... as a form of artistic expression to a modern generation.... In 1911 the English Folk Dance Society was formed and from that seed grew our present organization. The methods adopted were classes, country parties, festivals, demonstrations [Maud Karpeles, in Thompson 1953: 199ff.].

The majority of its members come to its dances, classes and festivals for recreation and enjoyment, and ... a small and serious body of people look to it as a focus for research [Williams 1958: 110–11].

As a result of these activities in England, a society was formed in the United States. At first it was a branch of the English Folk Dance Society, but later it became the autonomous Country Dance Society of America.

Similar societies have mushroomed throughout Europe, at first under scholarly guidance, but recently also without it. For instance, in Scandinavia "thousands and thousands have joined in the dance societies so that just now in Helsinki this summer we have a dancing festival with 1500 dancers from all of the northern parts of the country" (Otto Anderson, in Thompson 1953: 242–43).

That Switzerland has a wealth of old dances became known only slowly after 1930. Volumes of dances were published after that date [by Louise Witzig and others].... Now Switzerland is industrialized, but since 1935 various dance-groups have been formed.... The groups also make tours abroad. Thus the groups' repertoires contain both Swiss and foreign dances, though the latter cannot really be transplanted [Klenk 1954: 13].

The Catalan folklorist, Señor Capmany ... has worked to introduce Catalan dances and folk songs into schools, and has encouraged *Esbarts*, or societies for folk dance.... His solid work in *Folklore y Costumbres de España, El Baile y la Danza* ... was done in 1931, when traditional dances were left to the people of the soil which bred them, and had not suffered the dangerous invasion by hundreds of youth-groups with diverse foundations unconnected with either soil or dance [Alford 1949: 54–55].

American-born folk-dance groups, as well as immigrants to the United States, show their fondness for European forms. Lately they have become captivated by the sophisticated Balkan and Israeli rounds. These groups greatly surpass the "Amerindian" imitators in desire for accuracy, desire for expert advice, and an increasing demand for background knowledge. All depend for this knowledge largely on books and records, and provide a market for good publications. Some, such as the Philadelphia Folk Dance Center, co-operate with dance notators. (See Lattimore project.)

SCHOLAR AND SHOWMAN

The relative excellence of productions of traditional dances by non-native dancers depends on the integrity of the leaders, as well as on the quality of the members, of the dance group. At times the work of such groups leads to scholarly research by its members. Thus, after two decades as concert performer of Plains Indian dances, Reginald Laubin in 1959 undertook a research project on the history and cultural place of American Indian dances. Again, Bill Holm's interest in Kwakiutl dances has developed as follows (letter, Oct. 10, 1958):

Boas' *Social Organization and Secret Societies* is and has been our "Bible," supplemented by the material in *BAE* 35th Annual Report. It wasn't until about seven years ago that the opportunity came to see real Kwakiutl dancing, and from that time to the present we have been getting firsthand information at an ever increasing rate. Four years ago a group of us began building a Kwakiutl-style house. Two years ago we developed a really rewarding relationship with Mungo Martin and his family.

I always try to stick to the traditional form of the dance as I know it, subject to the limitations imposed by space, number of performers available, and the million other things that can make it difficult to reproduce a dance.

From performances given at his Jacob's Pillow University of the Dance, Shawn notes: "It is interesting to me to see that Shrimathi Gina [an American] gives a meticulously correct rendition of the various schools of Hindu dance, whereas Ram Gopal, a real Hindu, is more theatrical and free in his dances than Gina" (letter, July 5, 1958).

Josefina Garcia states frankly: "My work has been mostly ethnological instead of ethnic" (letter, July 28, 1958).

On the other hand, Paul Virsky, Director of the Ukrainian State Cossack Company, is quoted in Purdon (1958: 162) as saying that, "The future blending of folk dance and classic dancing ... is not, therefore, to present untouched folk dances of Ukraine, but to express by their means choreographic ideas in a theatrical form."

Again, Franziska Boas says that when she uses "any ethnographic material in my choreography, I use it from the present creative dance point of view and weave it into the choreography" (letter, July 30, 1958).

The scholar may function as performer, though more often he will counsel backstage. Thus Kurath and Etta-wageshik, in the course of an acculturation study, assisted the Ottawa in a production of a "Sun Ceremony." Ethnologists are evaluating the Pan-Indian powwow (Howard 1955; Schusky 1957).

By their stage performances, native and non-native professionals have performed a valuable public service (see Shawn [4]✩). Such artists adapt, and often change, the materials for theatrical purposes, hence face problems different from those of the choreographer, who emphasizes accurate recording.

PROSPECT

This completes the circle back to the question of definition and scope of dance ethnology, to the distinction between dance of the people and of the artist. Shall we accept Thurston's broad category of "folk dance" and admit the latest jazz steps, which, in Shomer's words, are "as timely as to-day's headlines in the newspapers, as interesting as the pulse of America" (1943: 5)? Shall we accept Sach's all-inclusive concept of dance ethnology? Franziska Boas proposes: "It would be interesting to trace relations between folk, ritual and twentieth century modern movement, particularly into the ritualistic aspects of the 'Modern Dance' of the twenties, and what has become of it now" (letter, July 30, 1958).

Any dichotomy between ethnic dance and art dance dissolves if one regards dance ethnology, not as a description or reproduction of a particular kind of dance, but as an approach toward, and a method of, eliciting the place of dance in human life—in a word, as a branch of anthropology.

To establish dance ethnology as a subdiscipline, ethnologists must accept choreology as a science, and choreologists must accept the scholarly responsibilities of being ethnologists. Let us hear the point of view of an ethnologist, James H. Howard (letter, Oct. 16, 1958; see also Shawn [1]☆):

The study of the dance, or choreology, is not only a legitimate but a very necessary part of the ethnological study. It is sometimes difficult for persons in our own culture, where the dance has been relegated to (1) performances of a spectacular nature by a few professional artists and to (2) a rather mechanical means by which members of one sex meet the other half of the population, to understand the importance of dance, as dance, to members of other cultures. Perhaps because of the relative unimportance of dance in Western European culture, most ethnographers ignore the dance completely, as something beneath their consideration, or content themselves with descriptions of the dance costume plus a few striking features (i. e. the "torture" feature of the Sun dance, the "shooting" dance of the Midewiwin). It is necessary, however, if we are attempting to describe a culture, to at least accord the dance the same importance as it is given by members of that culture. Lines of direction, characteristic steps and movements are culture traits and as such are capable of, and deserve, the same sort of treatment as other cultural elements. It would be well for each ethnographer, as part of his training, to learn some form of dance notation in order that he might record the dances of the group he is studying with the same accuracy he employs in recording kinship data, basket weaves, or irrigation techniques.

Again, William N. Fenton states in a letter (Oct. 29, 1959):

There is now evidence to demonstrate that music and the dance of primitive peoples can be analyzed and published and that these materials make very good tools for cultural and historical analysis and reconstruction. To linguistics is now added a second and third dimension of motor behavior that expresses the covert culture. These externalize the behavior systems that are manifest in language, the dance, and music and they used to baffle the field ethnologist who knew them as the most real expression of the culture but they frustrated him because they eluded his net. Now there are devices for recording these cultural expressions with some hope that they can be analyzed.

Teamwork is the best solution to problems of effecting comprehensive coverage and accurate analysis, given the increasing scope and specialized demands of dance research. This means teamwork between choreographers for standardization of analytic methods and for exchange of data, teamwork between choreographers and ethnographers for constructive application of these methods, and even teamwork between scientists and laymen, the public, the therapist, and the industrialist.

Only teamwork can cope with the widening scope and rapidly changing subject matter of dance ethnology. By the time that choreology is an established subdiscipline, the aborigine will have vanished, and we will have to assess the place of dance not only in mestizo culture but in our own materialistic, hybrid culture.

Notes

1. Space does not permit a historical survey of research activities. However, my brief comments have been augmented by some valuable information on Europe by Wolfram. See Wolfram,☆ the English of which is a free translation from his German.

2. Wolfram☆ mentioned a number of publications which are not readily available in the United States. They have been incorporated into the Bibliography below, identified by a star. Wolfram considers Müllenhoff (1871) to be a notable pioneer effort, and considers that Böhme (1886), though not based on personal observations, is a valuable compilation of early source materials.

3. McFeat☆ comments, "I disagree with the statement . . . regarding 'the formlessness of jitterbugging.' My impression, especially when attending student dances, is that jitterbugging, which has been in form for at least twenty years, is a very conservative term for what is now called 'rock and roll.'" But I would answer that "rock 'n roll" is derived from jitterbugging and is equally lacking in formations.

Comments

By ERNA GUNTHER☆

[1] This is an overwhelming paper in its scope and may stimulate further work or frighten anyone inclined to timidity. I would prefer a separation between the true ethnological field and the folk dance of the European and American cultures. I can see the relationship theoretically, but the students in these two fields have such totally different background and orientation that it is difficult to include all their needs and attitudes in a single study.

[2] The definitions of dance ethnology which are quoted need closer scrutiny. I prefer the one of Franziska Boas, which would appeal most to the anthropologist. Kurath's definition of ethno-choreography says really the same as Boas, but it should be a definition of dance ethnology rather than "choreography," since the latter is limited in general usage to the dance pattern and its execution rather than the cultural relations of the dance.

[3] In the great mass of categories for study, there is one absent which I consider of interest, namely, the relation between the dance, costumes, and the place of performance. For instance, a dance executed by firelight before a large group is often seen in silhouette and demands large dramatic gestures, whereas one done for an intimate group in good light can use facial expressions and minute hand movements. If the dancer must handle any part of his cos-

tume, it restricts his arm movements. The relationship of costume to movement in the choreography of a modern dance is very apparent, and leads one to realize that in ethnic dances two lines of development often meet in traditional clothing, on the one hand, and dance gestures and stance, on the other. How are they reconciled?

[4] I hope this study will be the impetus to more accurate field work by teams, as suggested, as well as a detailed appraisal and review of what is actually hidden in many ethnographic sources, that deal with the dance only casually.

By TED SHAWN[*]

[1] It seems to me that Kurath has done the most complete assembling of facts, and from all possible angles, on her subject. This is an extremely valuable survey, for it places in one article everything that anyone could ask about what knowledge has been accumulated, where it can be found, and evaluations of work done, and it sets out all of the fields of further and necessary research.

I am delighted that anthropologists are finally waking up to the importance of dance. And I feel that dissemination of Kurath's article will promote recognition of the absolute necessity of including study of ethnic dance in any complete dance curriculum.

[2] I am completely in accord with Kurath's ideas on the necessity of teamwork by the dancer-choreographer, trained anthropologist, photographer, sound recorder, and dance notator. Wanting to have my own eyes and ears undistracted during the five days of the Corrobboree given in my honor in Australia in 1947, I had a writer for the *Australian National Geographic Magazine* take the motion and still pictures, unsupervised. He shot only on climaxes —the fastest and most violent movements—whereas many of the slower passages were of equal significance.

[3] Kurath probably does not know of the amount of ethnic film I have in my possession: besides that on the Australian Aborigines, I have films taken during my tour (1925–26) in India, Java, Ceylon, Darjeeling (Tibetan lamas), Japan, and many other parts of Asia. I also have films of dancing in Bali, of a "Bee Dance" by Moros in the Philippines, of some Oklahoma Indians taken by a surgeon who had access to dances not seen by any other White man, etc. etc. All these are catalogued, and a copy of this list is available to Mrs. Kurath or to anyone else who wants it (write: Ted Shawn, Jacob's Pillow Dance Festival, Inc., Box 87, Lee, Mass., U.S.A.). These films will eventually be added to the Denishawn Collection in the Dance Archives of the New York Public Library.

[4] Also, she does not give full credit to some of the people who have done a lifetime of research on dance—Roger and Gloria Ernesti, who have specialized on the Indians of the Pacific Northwest, and Mme. La Meri (Russell Merriweather Hughes), whose Ethnologic Dance Institute was the finest clearinghouse for knowledge and information on ethnic dance that ever existed, but had to be closed for lack of funds.

[5] Ann Hutchinson, President of the Dance Notation Bureau of America and author of a textbook on Labanotation, agrees with me that the training of notators for the special purpose of working on a dance ethnology team is an admirable project and one which she will endeavor to promote. The most important fact to be faced is that a system of notation to give the greatest service must be *universally accepted*. Exchange of information freely between anthropologists and dance ethnologists can come about only if everyone, everywhere, accepts and uses one system, and Labanotation is the most complete and scientific system ever evolved. Even if a "shorthand" system is used in the field, the shorthand notes should be transcribed into Labanotation at the earliest possible moment, while memory is still fresh.

By RICHARD WOLFRAM[*]

Folk-dance revival in Europe received its earliest impetus from Sweden and South Germany. In Sweden a student organization, "Philochorus," was founded in 1880. The founder, Philochorus, drawing inspiration from a Swedish ballet master who arranged folklike dances for the Royal Opera, made field trips and presented dance arrangements on the stage. Besides these staged "national" dances, true folk dances persevered in Sweden and engaged the attention of the Folkdansens Vänner (1893) and other organizations. This movement stimulated research in Denmark and the founding in 1900 of the Danish Foreningen for Folkedansens Fremme. This organization, in turn, stimulated Cecil Sharp, who had already collected dances and songs in England and who then founded the English Folk Dance Society in 1911. Swedish developments also influenced North Germany and its Wandervogel and Jugendbewegung. On the other hand, the Trachtenvereine of South Germany arose independently in 1884, and soon extended their interest from costumes to dances, though always for show purposes.

Serious research scholars had to counteract the emphasis on the stage. Like Cecil Sharp, with his accurate field work, Raimund Zoder in Austria accumulated accurate notes from 1903 on, then Ernst Hamza in 1914, and then others. The scholarly preoccupation spread to Germany, which has produced many exact descriptions and analyses since 1920. France has produced some clean-cut work, such as Monique Decitre's "Dansez la France." Among Romance nations, the Spanish, Bretons, and Catalans have best preserved their traditions. The Basques have also. Italy and France tend to arrange their dances for show purposes. In Russia, Yugoslavia, and Rumania, the government tries to preserve traditions and to encourage research, as well as large ensembles for displays.

National Folk Festivals began in the 1920's in Scandinavia and England. The first all-European folk-dance festival took place in 1934, in Vienna. Out of this came the important festivals and congresses which took place in London in 1935, and in Stockholm in 1939. And out of these grew up the International Folk Music Council and subsequent conferences in Switzerland, Italy, U.S.A., Yugoslavia, England, France, Brazil, Norway, Germany, Denmark, Rumania, and Austria.

Selected Source Materials

Of the published items annotated below as containing useful bibliographies, see especially Lekis (1958) and Horak (1959). For an almost complete list of journals containing articles on dance, see W. Edson Richmond's *Annual Bibliography of Folklore (Journal of American Folklore Supplement* [April, 1959] Pt. 2: 19–22).

A compilation of theses and dissertations based on dance research is for sale by the National Section on Dance, American Association for Health, Physical Education and Recreation, 1201 Sixteenth St. N.W., Washington 16, D.C.

Films, recordings, notes, manuscripts, and rare books from various countries are on file in the libraries of the Dance Notation Bureaus of New York City (47 West 63 St.) and Philadelphia, Pennsylvania (271 South Van Pelt St.); the Jacob's Pillow University of the Dance (Lee, Massachusetts), The English Folk Dance and Song Society (Cecil Sharp House, 2 Regent's Park Road, London, N.W. 1), The Laban Art of Movement Centre (Addlestone, Surrey, England), the Manchester Folk Dance House (505

Wilbraham Road, Manchester, England), the Folkwangschule der Stadt Essen (22 A Essen-Werden, Germany), the Tanzarchiv of the Musikwissenschaftliches Institut in Regensburg, Germany, and other centers.

The addresses of persons holding private collections of any kind of material relevant to dance ethnology are available from the continually growing files of the Dance Research Center, 1125 Spring Street, Ann Arbor, Michigan, U.S.A.

Bibliography

AGARKAR, A. J. 1950. *Folk-dance of Maharashtra.* Bombay: Joshi.
AKIMOTO FUMI. 1953. Japanese dance. Unpublished M.A. thesis, Juilliard School of Music, New York. (Labanotation.)
ALFORD, VIOLET (Ed.). 1948–52. *Handbooks of European national dances.* London: Parrish; New York: Crown Publishers.
———. 1949. Don Aurelio Capmany. *Journal of the International Folk Music Council* 1:54–55.
ALFORD, VIOLET, and RODNEY GALLOP. 1935. *The traditional dance.* London: Methuen.
ALMASADZE, G. 1930. *Ansambly Azerbaijanskovo Narodnovo Tantsy.* Baku, Azerbaijan: Azerbaijan State Philharmonic Society.
ALMEIDA, RENATO. 1958a. O Bumba-Meu-Boi de Camassari. *Modulo* 2, No. 9:7–13. Rio de Janeiro.
———. 1958b. Brazilian folk ballets. *Folklorist* 4:145–48. Manchester, England.
AND, METIN. 1959. Dances of Anatolian Turkey. *Dance Perspectives* 1, No. 3.
ARNHEIM, RUDOLPH. MS. "Movement and the psychology of expression," in The Function of Dance in Human Society (Second Seminar) (ed. FRANZISKA BOAS), New York, 1946.
BALACI, EMANUELA, and ANDREI BUCSAN. 1956. Folclorul coregrafic din Sibiel. *Revista de Folclor* 1:213–47. Bucharest.
BARBEAU, MARIUS. 1957. *My life in recording Canadian Indian lore.* Ethnic Folkways Record FG 3502.
BARBEAU, MARIUS, et al. 1958. *Dansons à la ronde (roundelays).* Ottawa: National Museum of Canada.
BEALS, RALPH L. 1945. *The contemporary culture of the Cahita Indians.* Bureau of American Ethnology Bulletin 142.
BERDYCHOVA, JANA. 1951. *Tance Sovetskych Narodu.* Popis s obrazky a hudebnim doprovodem. Prague.
BLANCHARD, ROGER. 1943. *Les danses du Limousin.* Paris: Maisonneuve.
BOAS, FRANZ. 1944. "Dance and music in the life of the Northwest Coast Indians of North America (Kwakiutl)," in *The Function of Dance in Human Society (First Seminar)* (ed. FRANZISKA BOAS), pp. 7–18. New York: Author.
———. 1955. *Primitive art.* 1st ed., 1927. New York: Dover.
BOAS, FRANZISKA (Ed.). 1944. *The function of dance in human society (first seminar).* New York: Author.
———. MS. The function of dance in human society (second seminar), New York, 1946. (Typed copy provided by JOANN KEALIINOHOMOKU.)
BÖHME, FRANZ M. 1886. *Geschichte des Tanzes in Deutschland.* 2 vols. 339 pp. text, 221 pp. music. Leipzig.☆
BOWERS, FAUBION. 1953. *The dance in India.* New York: Columbia University Press.
BROWN, DONALD N. 1959. The dance of Taos Pueblo. Senior Honors Thesis, Harvard University, Cambridge, Mass.
CAMPBELL, JOSEPH. 1946. The ancient Hawaiian hula. *Dance Observer* 13, No. 3:32–33.
CAMPOBELLO, NELLIE, and GLORIA CAMPOBELLO. 1940. *Ritmos indigenas de México.* México, D.F.: Oficina Editora Popular.
CAVALLO-BOSSO, J. R. MS. Kumanche of the Zuni Indians of New Mexico. Thesis submitted at Wesleyan University for B.A., with distinction in Ethnomusicology, 1956.
CHILKOVSKY, NADIA. 1955–56. *Three R's for dancing.* (Series.) New York: Witmark.
———. 1958. Editorial. *Dance Notation Record* 9, No. 2:1.
———. 1959. *American bandstand dances in labanotation.* New York: Witmark.
CHUJOY, ANATOLE (Ed.). 1949. *Dance encyclopedia.* New York: A. S. Barnes.
COURLANDER, HAROLD. 1944. "Dance and dance-drama in Haiti," in *The Function of Dance in Human Society (First Seminar)* (ed. FRANZISKA BOAS), pp. 35–45. New York: Author.
CROSFIELD, DOMINI. 1948. *Dances of Greece.* (Handbook series.) London: Parrish.
CZARNOWSKI, LUCILLE. 1950. *Dances of early California days.* Palo Alto, Calif.: Pacific Books.
DEAN, BETH, and VICTOR CARELL. 1955. *Dust for the dancers.* New York: Philosophical Library.
DEMPSEY, HUGH A. 1956. Social dances of the Blood Indians of Alberta, Canada. *Journal of American Folklore* 69, No. 271:47–52.
DMITRI. 1958. Characteristics of Brazilian dance. *Dance Notation Record* 9, No. 2: 9–11.
DOZIER, EDWARD P. MS. The Rio Grande Pueblos (including analyses of dance types and organization), in seminar on Differential Culture Change, University of New Mexico, 1956.
DUGGAN, ANNA S., et al. 1948. *Folk dances of the United States and Mexico.* (From "Folk Dance Library" Series.) New York: A. S. Barnes. (Useful bibliographies.)
DUTTON, BERTHA P. 1955. *New Mexico Indians and their Arizona neighbors.* Santa Fe: New Mexico Association on Indian Affairs.
EBERLE, OSKAR. 1955. *Cenalora: Leben, Glaube, Tanz und Theater der Urvölker.* Switzerland: Walter Verlag. (Large bibliography.)
EVANS, BESSIE, and MAY G. EVANS. 1931. *American Indian dance steps.* New York: A. S. Barnes.
FENTON, WILLIAM N. 1941a. Tonawanda longhouse ceremonies ninety years after Lewis Morgan. Bureau of American Ethnology Bulletin 128:140–66.
———. 1941b. Masked medicine societies of the Iroquois. *Smithsonian Report for 1940,* pp. 397–430.
———. 1955. The maple and the passenger pigeon in Iroquois life. Albany: University of the State of New York.
FENTON, WILLIAM N., and G. P. KURATH. 1953. *The Iroquois eagle dance, an offshoot of the Calumet dance.* Bureau of American Ethnology Bulletin 156. (Large bibliography.)
FORTÚN DE FOLIN, JULIA ELENA. 1957. *La navidad en Bolivia.* La Paz: Ministerio de Educación. (Good bibliography.)
GAMBLE, JOHN I. 1952. "Changing patterns in Kiowa Indian dances," in *Acculturation in the Americas: Proceedings of the Twenty-Ninth International Congress of Americanists* (ed. SOL TAX), Vol. 2, pp. 94–104. Chicago: University of Chicago Press.

GARCIA, JOSEFINA. 1958. Latin-American dances. Unpublished Ph.D. dissertation, Texas Women's University, Denton, Tex.
GILLESPIE, JOHN D. MSa. Notes on the Shawnee bread dance.
———. MSb. Some Eastern Cherokee dances and songs.
GONYEY, SANDOR, and LASZLÓ LAJTHA. 1937. *Tánc.* Magyarság Szellemi Néprajza IV. Budapest.☆
GÖTLIND, J., and H. GRÜNER NIELSEN (Eds.). 1933. *Idrott och lek, Dans.* Nordisk Kultur XXIV. 198 pp. Stockholm, Oslo, Copenhagen. (Dance in Denmark, Norway, Faroes.)☆
GOMBERG, WILLIAM. MS. "Time and motion studies in modern industry," in The Function of Dance in Human Society (Second Seminar) (ed. FRANZISKA BOAS), New York, 1946.
GORER, GEOFFREY. 1944. "Function of dance forms in primitive African communities," in *The Function of Dance in Human Society (First Seminar)* (ed. FRANZISKA BOAS), pp. 19–34. New York: Author.
GUNTHER, ERNA. MS. A preliminary analysis of Kwakiutl dance.
HAMZA, ERNST. 1957. *Der Ländler.* Forschungen zur Landeskunde von Niederosterreich, Bd. 9. Vienna.
HARASYMCZUK, ROMAN W. 1939. *Tánce Huculskie.* 304 pp. text, 56 pp. music. Lwów, U.S.S.R.☆
HAYES, ELIZABETH R. 1955. *Dance composition and production.* New York.
HEIKEL, YNGVAR. 1939. *Folkdans.* Finlands Svenska Folkdiktning 6. 496 pp. Helsingfors: Svenska Litteratursällskapet i Finland.☆
HEIN, NORVIN. 1958. The Rām Lilā. *Journal of American Folklore* 71, No. 281:279–304.
HERMAN, MICHAEL. 1947. *Folk dances for all.* New York: Barnes and Noble.
HERSKOVITS, MELVILLE J. 1941. *The myth of the Negro past.* New York: Harpers.
———. 1949–50. "Possession," in *Dictionary of Folklore, Mythology and Legend* (ed. MARIA LEACH), Vol. 2, pp. 881–82. 2 vols. New York: Funk and Wagnalls.
HICKMANN, HANS. 1957. Danses de l'Egypte pharaonique et moderne. *Folklorist* 4:14–17.
HOERBURGER, FELIX. 1956. *Die Zwiefachen.* Berlin: Akademie Verlag.
HOLDEN, RICKEY, et al. 1956. *The contra dance book.* Newark, N. J.: American Square Dances.
HOLT, CLAIRE, and GREGORY BATESON. 1944. "Form and function of the dance in Bali," in *The Function of Dance in Human Society (First Seminar)* (ed. FRANZISKA BOAS), pp. 46–52. New York: Author.
HORAK, KARL. 1948. *Schuhplattlerschlüssel: Beiträge zur Volkskunde Tirols.* Festschrift zu Ehren Hermann Wopfners, Schlern-Schriften, Bd. 53. Innsbruck.
———. 1959. *Bibliographie des Volkstanzes in Österreich.* Innsbruck: Author.
HORST, LOUIS. 1940. *Pre-classic dance forms.* New York: Dance Observer.
HOWARD, JAMES H. 1955. Pan-Indian culture of Oklahoma. *Scientific Monthly* 81, No. 5:215–20.
———. MS. The Turtle Mountain Plains-Ojibway. (Sections on dance organizations and forms, and partly transcribed song tape, by G. KURATH.)
HOWARD, JAMES H., and G. KURATH. 1959. Ponca dances, ceremonies and music. *Ethnomusicology* 3, No. 1:1–14.
HUTCHINSON, ANN. 1954. *Labanotation.* New York: New Directions.
JANKOVIC, LJUBICA S., and DANICA S. JANKOVIC. 1934–51. *Folk dances 1–6, a summary*

of *Narodne Igre*. Belgrad: Council of Science and Culture.
———. 1952a. *Narodne Igre*. Vol. 7. Belgrad.
———. 1952b. *Dances of Yugoslavia*. (Handbook series.) London: Parrish.
———. 1955. Provilno u Nepravilnome (Le Regulier dans l'Irregulier). *Zvuk*. *Jugoslavenska Muzicka Revija* 2–3:65–79.
———. 1957. A contribution to the study of the survival of ritual dances. Serbian Academy of Sciences Monographs, Vol. 271. Belgrad.
JONES, A. M. 1953. Folk music in Africa. *Journal of the International Folk Music Council* 5:36–39.
KADMAN, GURIT. 1956. The folk dance of Israel. *Folk Dancer* 3, No. 1:165–67. Manchester, England.
KAPLAN, SHIRLEY, MS. Notes on folk-dances of India. (Film and tape.)
KAWANO HIROMICHI. 1956. *Ainu Odori*. Tokyo: Nirei Shobo.
KEALIINOHOMOKU, JOANN. 1958. A comparative study of dance as a constellation of motor behaviors among African and United States Negroes. Unpublished M.A. thesis. Northwestern University, Evanston Ill. (Analysis of Africa from films, of America from observation; dance notation.)
KENNEDY, DOUGLAS. 1949. *England's dances*. London: Bell.
KIRSTEIN, LINCOLN. 1935. *Dance: A short history of classic theatrical dancing*. New York: Putnam.
KLEIN, ERNST. 1937. Om Polskedanser. *Svenska Kulturbilder*, n.f., Bd. 5, Div. IX, X. Stockholm.☆
KLENK, KARL. 1954. Der Volkstanz in der Schweiz. *Folk Dancer* 1, No. 5:9–13.
KNUST, ALBRECHT. 1956a. Letter to the editor. *Dance Notation Record* 6, No. 5–6:13.
———. 1956b. *Abriss der Kinetographie Laban*. Hamburg: Das Tanzarchiv.
———. 1958. Letter to the editor. *Dance Notation Record* 9, No. 1:3–4.
KNUST, ALBRECHT, with the assistance of DIANA BADDELEY and VALERIE PRESTON. 1958. *Handbook of kinetography Laban*. (English version of 1956b.) 2 vols. Hamburg: Das Tanzarchiv.
KONYVKIADO, MUVELT. 1954. *Hungarian dances*. (Series Neptancosok.) Budapest.
KURATH, GERTRUDE P. 1949. Mexican Moriscas: A problem in dance acculturation. *Journal of American Folklore* 62, No. 245:67–106.
———. 1949–50. "Dance, folk and primitive" and other entries, in *Dictionary of Folklore, Mythology and Legend* (ed. MARIA LEACH), pp. 276–96, passim. 2 vols. New York: Funk and Wagnalls.
———. 1951. Local diversity in Iroquois dance and music. Bureau of American Ethnology Bulletin 149, Pt. 6:109–37.
———. 1952. "Dance acculturation," in *Heritage of Conquest* (ed. SOL TAX), pp. 233–42. Glencoe, Ill.: Free Press.
———. 1953. Native choreographic areas of North America. *American Anthropologist*, n.s., 55:153–62.
———. 1954. The Tutelo harvest rite. *Scientific Monthly* 76, No. 3:87–105.
———. 1956a. Dance relatives of mid-Europe and middle America. *Journal of American Folklore* 69, No. 273:286–92.
———. 1956b. Antiphonal songs of Eastern Woodland Indians. *Musical Quarterly* 42:520–26.
———. 1957a. Algonquian ceremonialism and natural resources of the Great Lakes. Reprint 22. Bangalore: Indian Institute of Culture.
———. 1957b. Basic techniques of Amerindian dance. *Dance Notation Record* 8, No. 4:2–8.
———. 1957c. Notation of a Pueblo Indian corn dance. *Dance Notation Record* 8, No. 4:9–11.
———. 1958a. Plaza circuits of Tewa Indian dancers. *El Palacio* 65, No. 1–2:11–26.
———. 1958b. Game animal dances of the Rio Grande. *Southwestern Journal of Anthropology* 14:438–48.
———. 1959. Menomini Indian dance songs in a changing culture. *Midwest Folklore* 9, No. 1:31–38.
———. MSa. Seneca song and dance style. 1951. American Philosophical Society Library. (Dance notation.)
———. MSb. Onondaga ritualism. 1952. New York State Museum Educational Library. (Dance notation.)
———. MSc. Tutelo dances and songs. 1953. (Dance notation.)
———. MSd. Animal rites, dances and songs of Eastern Woodland Indians. 1955. (Dance notation.)
———. MSe. Tewa Indian dance and music. 1957–58. (Dance notation; film and tape.)
KURATH, GERTRUDE P., and NADIA CHILKOVSKY. 1960. Jazz choreology. Proceedings of the Fifth International Congress of Anthropological and Ethnological Sciences (ed. ANTHONY WALLACE). Philadelphia: University of Pennsylvania Press.
KURATH, GERTRUDE P. and JANE W. ETTAWAGESHIK. MS. Religious customs of modern Michigan Algonquians. 1955. American Philosophical Society Library. (Dance notation.)
LABAN, JUANA DE. 1954. Movement notation: its significance to the folklorist. *Journal of American Folklore* 67, No. 265:291–95.
———. MS. Notes on Hungarian dances. (Labanotation).
LABAN, RUDOLF. 1928. *Kinetographie*. 2 vols. Vienna.
LABAN, RUDOLF, and F. C. LAWRENCE. 1947. *Effort*. London: MacDonald and Evans.
LAMBERT, OMER. n.d. *Danses Canadiennes*. Quebec: Imprimerie Nacionale.
LA MERI. 1948. *Spanish dancing*. New York: A. S. Barnes.
LANGE, CHARLES H. 1953. Economics in Cochiti culture change. *American Anthropologist* 55:174–94.
———. 1957. The Tablita, or corn dance of the Rio Grande Pueblo Indians. *Texas Journal of Science* 9, No. 1:59–74.
———. MS. The Pueblo of Cochiti, New Mexico: past and present. Austin: University of Texas Press. In press. (Chapters on ceremonialism, with dance and music analysis by G. KURATH.)
LAPSON, DVORA. 1954. *Dances of the Jewish people*. New York: Jewish Education Committee of New York.
LATTIMORE, ALICE. 1957. Kalamatianos. *Dance Notation Record* 8, No. 1:8.
LAWSON, JOAN. 1953. *European folk dances*. Toronto and London: Pitman.
LAWTON, SHAILER U. MS. Physiological determinants in movement, in "The Function of Dance in Human Society (Second Seminar) (ed. FRANZISKA BOAS), New York, 1946.
LEKIS, LISA. 1956. Origin and development of ethnic Caribbean dance and music. Unpublished Ph.D. dissertation, University of Florida, Gainesville, Fla.
———. 1958. *Folk dances of Latin America*. New York: Scarecrow Press.
———. MS. Notes on Peruvian and Ecuadorian dances.
LORING, EUGENE, and A. J. CANNA. 1956. *Kinetography*. Los Angeles: Author.
LUGOSSY, EMMA, and SANDOR GONYEY. 1947. *Magyar Nepi Tancok*. Budapest.

MANN, JOHN. MS. Teaching and research in ethnic dance. *Ethnomusicology*. In press.
MANSFIELD, PORTIA. 1952. The Conchero dancers of Mexico. Unpublished Ph.D. dissertation, New York University, New York, N.Y. (Music by RAOUL GUERRERO.)
MARIÁN, RETHEI PRIKKEL. 1924. A Magyasag Tancai. 311 pp. Budapest.☆
MARTI, SAMUEL. 1959. *Danza precortesiana*. Sobretiro de Cuadrenos Americanos, No. 5 de 1959.
———. MS. *Canto, música y danza precortesiana*. México, D.F.: Fondo de Cultura Económico and Instituto Nacional de Antropología e Historia. In press.
MASON, BERNARD S. 1944. *Dances and stories of the American Indian*. New York: A. S. Barnes.
MAYO, MARGOT. 1948. *The American square dance*. New York: Sentinel Books.
MATIDA KASYO. 1938. *Odori (Japanese dance)*. Tokyo: Board of Tourist Industry, Japanese Government Railways.
MCALLESTER, DAVID P. MS. Comanche sign language. Paper in Ethnomusicology, at Columbia University, 1941.
MEAD, MARGARET. 1949. *Coming of age in Samoa*. New York: Mentor Books. First published in 1928.
———. MS. "Dance as an expression of culture patterns," in The Function of Dance in Human Society (Second Seminar) (ed. FRANZISKA BOAS), New York, 1946.
MOERDOWO, R. 1957. Dances in Indonesia. *Folklorist* 4:63–66.
MOISEYEV, IGOR. 1951. *Tänze der Völker der Sowietunion*. Berlin.
———. 1952. *Orosz Heptsancszvit*. Budapest.
MOONEY, GERTRUDE X. 1957. *Mexican folk dances for American schools*. Coral Gables, Fla.: University of Miami Press.
MÜLLENHOFF, K. 1871 (and later supplements). *Über den Schwerttanz*. Berlin.☆
MURDOCK, GEORGE P., et al. 1954. *Guia para la clasificación de los datos culturales*. New Haven and Washington, D.C.: Pan-American Union.
NIELSEN, H. G. 1917. *Vore äldste Folkedanse*. 71 pp. text, 24 pp. music.☆
———. 1920. *Folkelig Vals*. 83 pp. text, 47 pp. music. Kopenhagen.☆
OKUNEVA, B. B. 1951. *Russkie Narodnye Chora i Tantsy*. Sophia.
ONEGINA, T. D. 1938. *Moldavsky Tanets*. Moscow.
POLLENZ, PHILIPPA. 1946. Some problems in the notation of Seneca dances. Unpublished M.A. thesis, Columbia University, New York. (Labanotation.)
———. 1949. Methods for the comparative study of the dance. *American Anthropologist*, n.s., 56:428–35.
PROCA, VERA. 1956. Despre notarea dansului popular rominesc. *Revista de Folclor* 2, No. 1–2:135–71. Bucharest.
———. 1957. Despre notarea dansului popular rominesc. *Revista de Folclor* 2, No. 1–2:65–92.
PROKOSCH, GERTRUDE. 1938. Rhythms of work and play. *Journal of Health and Physical Education* 9:294–97.
PURDON, M. E. 1958. Ukrainian Cossack Company. *Folklorist* 4, No. 6:158–62.
RHODES, WILLARD. 1956. On the subject of dance. *Ethnomusicology Newsletter* 7:1–9.
ROGERS, HELEN P. 1958. A complete record for the dance. *Dance Notation Record* 9, No. 1:5–6.
SACHS, CURT. 1933. *Eine Weltgeschichte des*

Tanzes. Berlin: Reimer. (Huge bibliography.)
———. 1953. *Rhythm and tempo*. New York: Norton.
SALZ, BEATE. 1955. *The human element in industrialization*. Memoir, American Anthropological Association, No. 85.
SANTA ANA, HIGINIO VASQUEZ. 1940. *Fiestas y costumbres Méxicanas*. México, D.F.: Ediciones Botas.
SCHMITZ, CARL A. 1955. *Balam: Der Tanz- und Kultplatz in Melanesien*. Emsdetten, Westphalia: Lechte Verlag.
SCHUSKY, ERNEST. 1957. Pan-Indianism in the eastern United States. *Anthropology Tomorrow* 6:116–23. Chicago: University of Chicago Anthropology Club.
SEDILLO-B., MELA. 1935. *Mexican and New Mexican folk dances*. Albuquerque: University of New Mexico Press.
SHAWN, TED. 1929. *Gods who dance*. New York: Dutton. (Valuable bibliography.)
SHOMER, LOUIS. 1943. *Swing steps*. New York: Padell Book Co.
SINGER, MILTON B. 1958. The great tradition in a metropolitan center: Madras. *Journal of American Folklore* 71, No. 281:347–88.
SLOTKIN, J. S. 1955. An intertribal dancing contest. *Journal of American Folklore* 78, No. 268:224–28.
———. 1957. *The Menomini powwow*. Milwaukee Public Museum Publications in Anthropology No. 4.
SOLARI, MARIA LUISA. 1958. Notación de la danza. *Revista Musical Chilena* 12, No. 58:42–58.
SPECK, FRANK G. 1949. *Midwinter rites of the Cayuga longhouse*. Philadelphia: University of Pennsylvania Press.
SPECK, FRANK G. and LEONARD BROOM. 1951. *Cherokee dance and drama*. Berkeley and Los Angeles: University of California Press.
SPENCE, LEWIS. 1947. *Myth and ritual in dance, game and rhyme*. London: Watts.
SPREEN, HILDEGARD L. 1949. *Folk-dances of South India*. New York: Oxford University Press.
STURTEVANT, WILLIAM C. 1954. The medicine bundle and busks of the Florida Seminole. *Florida Anthropologist* 7, No. 2:31–70. (Bibliography.)
Tantsi Narodov SSSR. ("Folk Dances of the USSR.") 1955. Moscow: Iskusstvo.
TAX, SOL. 1952. "Economy and technology," in *Heritage of Conquest* (ed. SOL TAX), pp. 43–75. Glencoe, Ill.: Free Press.
THOMPSON, STITH (Ed.). 1953. *Four symposia on folklore*. Bloomington: Indiana University Press.
THUREN, HJALMAR. 1908. *Folkesangen paa Färöerne*. Folklore Fellows Publications, Northern Series 2. 337 pp. Kopenhagen. (Over 100 songs.)☆
THURSTON, HUGH A. 1954a. *Scotland's dances*. London: G. Bell.
———. 1954b. What is a folk dance? *Folk Dancer* 1, No. 1:4–6.
TITIEV, MISCHA. 1955. *The science of man*. New York.
TOLMAN, BETH, and RALPH PAGE. 1937. *The country dance book*. New York: A. S. Barnes.
TOMKINS, WILLIAM. 1929. *Universal Indian sign language*. San Diego, Calif.: Author.
TSOUKALAS, NICHOLAS. 1956. *Spanish dancing: zapateado movements*. Detroit: Author.
TULA. 1952. Therapeutic dance rhythms. *Dance Observer* 19, No. 10:117–18.
TURLEY, FRANK. MS. The present-day Oklahoma fancy war dance. 1959.
VEGA, CARLOS. 1952. *Las danzas populares argentinas*, Vol. 1. Buenos Aires: Ministerio de Educación de la Nación. (Comprehensive bibliography.)
VENABLE, LUCY, and FRED BERK. 1959. *Ten folk dances in labanotation*. New York: Witmark.
VISKI, KÁROLY. 1937. *Hungarian dances*. 193 pp. text, 34 pls. London and Budapest.☆
VYCPALEK, JOSEF. MS. Česke Tance. Czechoslovakia. In press.☆
WELTFISH, GENE. MS. "Patterns of work in primitive industry," in The Function of Dance in Human Society (Second Seminar) (ed. FRANZISKA BOAS), New York, 1946.
WILDER, STAFFORD. 1940. The Yaqui deer dance. Unpublished M.A. thesis, University of Arizona, Tucson, Ariz. (Diagrams.)
WILLIAMS, RALPH VAUGHAN. 1958. The English folk dance and song society. *Ethnomusicology* 2, No. 3:108–12.
WISSLER, CLARK. 1916. General discussion of shamanistic and dancing societies. *Anthropological Papers of the American Museum of Natural History*, Vol. 2, Pt. 12.
WOLFRAM, RICHARD. 1931. Volkstanz—nur gesunkenes Kulturgut? *Zeitschrift für Volkskunde*. Berlin.☆
———. 1933. Die Frühform des Ländlers. *Zeitschrift für Volkskunde*. Berlin.☆
———. 1934. Altersklassen und Männerbünde in Rumanien. *Mitteilungen der Anthropologische Gesellschaft in Wien* 64:112–28.
———. 1937. Deutsche Volkstänze. *Bilder der deutschen Volkskunde* (ed. A. SPAMER), Vol. 5. 49 pp. text, 44 pls. Leipzig, Germany.☆
———. 1951. *Die Volkstänze in Österreich und verwandte Tänze iy Europa*. Salzburg: Müller. (Large bibliography.)
———. 1954. Der Schwerttanz. *Helleiner Heimatbuch*. (Series Heimat Österreich.) Graz, Drechsler.
———. 1956. European song-dance forms. *Journal of the International Folk Music Council* 8:32–35.
———. MS. Schwerttanze Europas. 800 pp. Three fascicles appeared Bärenreiterverlag, Kassel. (Diagrams of 30 dances.)
ZHORNITZKAYA, M. YA. 1956. *Yakutskie Tantsy*. Yakutsk, Siberia.
ZODER, RAIMUND. 1938. Der Deutsche Volkstanz. *Deutsches Volkstum* 3:137–84. Berlin.
———. 1950. *Volkslied, volkstanz und volksbrauch in Österreich*. Wien: Doblinger.

Projects

CASTNER, RICHARD L. A history of traditional American country dancing, to 1800.
CHILKOVSKY, NADIA. Observation and notation of national traits in effort behavior. (Laban Effort symbols.)
CHILKOVSKY, NADIA, and Students of Philadelphia Musical Academy and Philadelphia Dance Academy. Notes on regional jazz dances. (Labanotation.)
DELZA, SOPHIA. Notes on Chinese dances. (Labanotation.)
GARCIA, JOSEFINA. Labanotation of Latin American dances.
HOERBURGER, FELIX. Bibliography and critical inventory of folk dance publications in Europe.
HOLM, WILLIAM. Reconstruction of Kwakiutl dances.
KURATH, GERTRUDE P., and Dance Research Center Associates. Ritual Drama of the American Indian. (For *Dance Perspectives*).
LATTIMORE, ALICE. Notes on mudras of India, on Balkan, Slavic, Greek and other folk dances. (Labanotation.)
LAUBIN, REGINALD, and GLADYS LAUBIN. A historical study of American Indian dances.
MARTI, SAMUEL, and G. KURATH. Manual of dance analysis. (For SAMUEL MARTI MS.)
THURSTON, HUGH A. Dictionary of folk dance terms.
WEST, LA MONT. American Indian sign language. Indiana University thesis.

Non-Commercial Films

BOAS, FRANZ. Kwakiutl dances. University of Washington.
CHILKOVSKY, NADIA and LEAH DILLON. Complete vocabulary of Hindu mudras. Philadelphia Dance Academy.
DEAN, GETH, and VICTOR CARELL. Australian dances.
DEHN, MURA. History of Jazz Dance.
ENGLISH FOLK DANCE AND SONG SOCIETY. English dances. Cecil Sharp House.
FIRST, GEORGIA. Dances of Fox Indians. (Also slides.)
GERSON-KIWI, EDITH. Wedding ceremonies and dances of Kurdistan Jews.
IRISH FOLKLORE COMMISSION. Mummers and other Irish dances.
KAPLAN, SHIRLEY WIMMER. Folk dances of India.
(a) Twenty-one dances during Republic Day Celebration, New Delhi, Jan. 1957. Partial notation of nine dances, from Assam, Himachal Pradesh, Kashmir, Kerala, Orissa, Uttar Pradesh. (600 ft. color; taped music expected from Sangeet Natak Akadami.)
(b) Kandyan dance in Ceylon. (300 ft.; tape of drum and voice.)
(c) Girls' dance from Ellora. (100 ft.; no tape.)
KURATH, GERTRUDE P. Edited tapes and slide series at Dance Research Center, inter alia:
(a) Michigan Indian dances. (150 ft.; tapes.)
(b) Menomini dances. (100 ft.; tape.)
(c) Santa Clara plaza dances. (150 ft.; tape.) Copy at Wenner-Gren Foundation.
(d) Jazz dances of White and Negro teen-agers of Ann Arbor. (100 ft.)
MANSFIELD, PORTIA. Edited films. Copies, Perry-Mansfield School of Theater.
(a) The Conchero dances of Mexico. Sound, music arr. by RAOUL GUERRERO. (Two reels: Background [10 min.], Ceremonies [18 min.])
(b) Cowboy squares. (12 min.)
(c) Southwest Indian dances (Tewa Pueblos, Acoma, Cochiti, Taos). (12 min.)
(d) Nepalese dances. (Not edited.)
MERRIAM, ALAN P. Ekonda, Africa. (Edited.)
NICHOLLS, THAD. Yaqui Easter fiesta, Pascua, Arizona. (Edited.)
PETER, ILKA, and HERBERT LAGER. Austrian "Trestern."
PINKERSON, FRANCES. Ceremonial dances of Mexico (Volador, Moros y Cristianos, Apaches, Pastorcitas, etc.). (400 ft.; edited.)
POSPISIL, F. German Sword Dances.
RAYE, FLORIS. Hawaiian dances. 40 King's Road, London.
SHAWN, TED [3]☆ Dances of India, Java, Ceylon, Darjeeling, Tibet, Japan (1925–26), Australian Corroboree (1947), Bali, Philippine Moros, Oklahoma Indians, and others. List, Jacob's Pillow University of the Dance, Lee, Mass.
STREHLOW, THEODOR G. H. Aranda and Loritja totemistic ceremonies, Australia.
TURNER, ALONSO. West African journey.
WOLFRAM, R. Norwegian "Springar" (1939).

A System for the Notation of Proxemic Behavior

EDWARD T. HALL

A System for the Notation of Proxemic Behavior[1]

EDWARD T. HALL
Illinois Institute of Technology

INTRODUCTION

THIS is one of a series of papers on Proxemics,[2] the study of how man unconsciously structures microspace—the distance between men in the conduct of daily transactions, the organization of space in his houses and buildings, and ultimately the layout of his towns.

The aim of this paper is to present a simple system of observation and notation with a view to standardizing the reporting of a narrow range of microcultural events. The system is far from perfect; but if it directs attention to certain behavior, it will have achieved its purpose. However, before proceeding to the descriptive portion of this paper, certain theoretical matters have to be dealt with.

The writer began systematic observations in a proxemic frame of reference when it became apparent that people from different cultures interacting with each other could not be counted on to attach identical meanings to the same or similar measured distances between them (Hall 1955, 1959, 1963, 1964). What was close to an American might be distant to an Arab.

Without a systematic observational and recording technique for such encounters, pinpointing interferences[3] is a slow, somewhat uncertain procedure requiring highly developed observational skills. Not all observers are equally skilled.

Levels of Awareness

Any culture characteristically produces a simultaneous array of patterned behavior on several different levels of awareness. It is therefore important to specify which levels of awareness one is describing.

Unlike much of the traditional subject matter of anthropological observation, proxemic patterns, once learned, are maintained largely out of conscious awareness and thus have to be investigated without resort to probing the conscious minds of one's subjects. Direct questioning will yield few if any of the significant variables, as it will with kinship and house type, for example. In proxemics one is dealing with phenomena akin to tone of voice, or even stress and pitch in the English language. Being built into the language, these are hard for the speaker to consciously manipulate.

Values of Proxemic Study

Indeed, the very absence of conscious distortion is one of the principal reasons for investigating behavior on this level,[4] for any step taken to eliminate a subject's conscious manipulation of the facade[5] presented to the world is desirable.[6] Boas (1911) stressed this same point as the principal reason for integrating linguistics and ethnological research. There are, however, other reasons for studying proxemic behavior. Why is it, for example, that an American[7] who is approached too closely by a foreigner will feel annoyed? Why is it that the discomfiture often fails to pass when he gets to know the culture better, in spite of conscious striving to suppress these feelings? Why do these interferences commonly last a lifetime, and why do people take this sort of interference so personally? Why is there so little they apparently can do to relieve their feelings? One subject (an anthropological colleague), after 12 years of working with the French, still couldn't stand being approached as closely as Frenchmen normally do in conversation and used to barricade himself behind his desk, because he felt the French were still getting too familiar.

Considering the architect's persisting preoccupation with space, why is it that 2400 years since the building of the Parthenon Western man still lacks a method for noting and describing the experience of space?[8] Some of the answers to these paradoxes can be derived from the art and literature of our own culture.[9]

Acute observers from other fields—often neglected or ignored—also provide the anthropologist with helpful cues.

Significant Work in Related Fields

Hediger (1955), the Swiss ethologist, pioneered in the systematic observation of distance in vertebrates. He distinguishes between contact and non-contact species and was the first to describe personal and social distance in animals. His work continues to be of interest to anthropologists (Hediger 1961).

Dorner (1958) gives structure to the artist's continuous quest—from the neolithic to the present—to discover new and more satisfying means for portraying spatial relationships. From him we get a new slant on how man has organized and re-organized his actual perceptions.

Lynch (1960), after interviewing inhabitants of three major U. S. cities (Boston, Jersey City and Los Angeles), identified five elements intrinsic to the image of the city: paths, districts, edges, landmarks and nodes. These represented the subjects' own categories.

Grosser (1951), addressing himself solely to intimate, personal, and social distance, describes how and why portraits in the Western world are painted at certain specific distances. The distance employed by an artist when he paints a human subject is designed to communicate specific features of the personality and at the same time screen out all other features. Grosser pins his observations down to feet and inches.

Among the psychologists, the transactional group (Kilpatrick 1961) has made particular progress in isolating the principal means by which people judge the relationship and distances of objects, and in so doing has provided insight into how man unconsciously participates in the molding of his own perceptions. Gibson (1950) goes a long way in explaining how the total visual process stabilizes and synthesizes the ever-changing mosaic of images cast on the retina, converting them into a solid visual world.[10]

Barker and Wright (1954) have contributed significantly to the study of human ecology through their identifications of what they term standing behavior settings and objects (frames) for analyzing and describing the behavior of a community. These frames can be further analyzed in terms of constituent parts and categories that are subsumed under each heading. Barker and Barker (1961) discovered that attempts to impose categories of "behavior settings" on their English subjects were much less productive than deriving the categories from the subject's own preferences, and also that once established, these behavior setting categories tended to be self-perpetuating.

The combined insights of these writers—plus many more who could be cited—still leave several questions unanswered. By what means other than visual do people make spatial distinctions? How do they maintain such uniform distances from each other? And how do they teach these distances to the young?

METHODOLOGICAL CONSIDERATIONS

Foreign students studying in the United States comprised one group of our proxemic research subjects.

An unanticipated consequence of these interviews was added insight into what Goffman (1957) terms "the stuff of encounters" which was highlighted whenever there was interference between two patterns, or a perceived absence of patterning, during an encounter. These subjects reported suffering repeated alienations in encounters with Americans. There are many forms of alienation, but a frequent variety found was that which has been termed *lack of involvement*. Misinterpretation of American responses was traced to differences in the definition of what constituted proper listening behavior, some of which centered on use of the eyes. Further investigation revealed that there was also a virtual absence of skill in reading the minor cues as to what was going on behind the American facade (Goffman 1957).

In a broader context involving older subjects, Arabs complained of experiencing alienation particularly when interacting with that segment of the U. S. population which can be classed as non-contact (predominantly of North European origin, where touching strangers and casual acquaintances is circumscribed with numerous proscriptions). When approached too closely, Americans removed themselves to a position which turned out to be outside the olfactory zone (to be inside was much too intimate for the Americans). Arabs also experienced alienation traceable to a "suspiciously" low level of the voice, the directing of the breath away from the face, and a much reduced visual contact. Two common forms of alienation reported by American subjects were *self-consciousness at the cost of involvement* and *other-consciousness*. Americans were not only aware of uncomfortable feelings, but the intensity and the intimacy of the encounter with Arabs was likely to be anxiety provoking. The Arab look, touch, voice level, the warm moisture of his breath, the penetrating stare of his eyes, all proved to be disturbing. The reason for these feelings lay in part in the fact that the relationship *was not defined as intimate*, and the behavior was such that in the American culture is only permissible on a non-public basis with a person of the opposite sex.

In a different cultural setting, a Chinese experienced alienation during an interview when he was faced directly and seated on the opposite side of a desk, for this was defined as *being on trial*.

Research in proxemics has been restricted to culturally-specific behavior and it does not encompass other environmental or personality variables, such as noise level, temperature, and personality variables, all of which are important.[11] There are, nevertheless, a number of conceptual tools the anthropologist has at his disposal.

For example, the anthropologist knows that in spite of their *apparent* complexity, cultural systems are so organized that their content can be learned and controlled by all *normal* members of the group. Anything that can be learned has structure and can ultimately be analyzed and described. The anthropologist also knows that what he is looking for are patterned distinctions that transcend individual differences and are closely integrated into the social matrix in which they occur.

Proper observation can tease from the data the patterning that man gives to a behavioral system in order that he may use it and transmit it to others. The notation system which is given in this paper will help the field worker to focus his observations in such a way as to clarify the various structure points of a given proxemic system.

Only the notation system itself is given here. The history of how it was developed will be treated at length in later publications, along with the rather complex matter of arriving at definitions of such systems. The process of making explicit the rules that combine isolates into sets and patterns (Hall 1959) will be treated elsewhere.

In making observations of the sort required, and devising the notation system described below, this investigator owes a debt to several disciplines. Included in these are descriptive linguistics[12] and kinesics (Birdwhistell 1952), ethology, and the various psychoanalytic schools starting with Freud and ending with Fromm and Harry Stack Sullivan.[13] Even the writer's experience as a weather observer in World War II taught him the extremely useful nature of numerical codes in which information is associated not only with a numerical value but also with a position in the code. Weather codes were learned by thousands of men in a matter of weeks.

PROXEMIC NOTATION SYSTEM

Proxemic behavior can be seen as a function of eight different "dimensions" with their appropriate scales. Complex as proxemic behavior is in the aggregate, by proceeding one at a time, these dimensions can be recorded quickly and simply in the following order:

1) **postural—sex identifiers**
2) **sociofugal—sociopetal orientation (SFP axis)**
3) **kinesthetic factors**

4) touch code
5) retinal combinations
6) thermal code
7) olfaction code
8) voice loudness scale

Given the present meager state of our knowledge, a total of eight classes of events is sufficient to describe the distances (and the means of determining distances) employed by man. The systems are bio-basic, rooted in the physiology of the organism (Hall 1959; Hall and Trager 1953). With slight modifications, this system could also be used to describe distance behavior of other mammals. It conforms to Wallace's number eight given for basic building blocks of cultural systems (Wallace 1961). It is consistent with the criteria set by Goodenough in which observed units are ordered and interrelated according to contrasts inherent in the data (Goodenough 1956, 1957; Frake 1962).[14]

Each factor complex comprises a closed behavioral system which can be observed, recorded, and analyzed in its own right. On the proxemic level, however, each system is treated as a complex of isolates that will result in proxemes which combine into sets and patterns in larger systems.[15] For example, there are almost endless variations on posture. One observer (Hewes 1955) records empirically some ninety positions. However, it is not necessary to note all variations. For example, in proxemic notation it is important to record only the sex of the subjects and whether they are standing, sitting or squatting, or lying down (prone).

Not all of the eight factors are of equal complexity, nor do all of them function at all times. The thermal and olfaction inputs are present only at close distances. Vision is more complex than either of these, and it is normally screened out only at very close distances.[16]

In the interest of speeding recording and with a view to future application of factorial analysis and computers to field records, every attempt has been made to be parsimonious. At each of the eight steps it is necessary for the observer to make only a few discriminations, in no case more than seven at one time. The present version of the system enables recording in 30–60 seconds; with practice, familiarity, and improvements, it should be possible to reduce the recording time to as little as 10–15 seconds.[17]

The notation system which follows is designed to systematize observation in the simplest possible way and at the same time to provide a record so that similar events can be compared across time and space.

Postural-sex Identifiers

One of the most essential operations in proxemic notation is to determine the sex and basic posture of the two individuals (whether they are standing, sitting or squatting, or prone). These distinctions are kept to a minimum in order to maintain congruity throughout the proxemic analytic level. On other analytic levels much greater detail has been observed and noted. For example,

R. L. Birdwhistell, who named and developed the field of kinesics and worked in close cooperation with descriptive linguists, including McQuown and Hockett, notes an extraordinary number of events (Birdwhistell 1952; McQuown 1957; Hockett 1958). The most recent number given[18] was over 30 different lines of recordings, each representing a possible analytic level.

Sex and basic posture can be noted by any one of three systems depending on the needs of the investigator: a pictographic mnemonic (iconic) symbol, a syllabic mnemonic, or simple number code. These are:

man prone	⌐▵	m/pr	1
man sitting or squatting	⌐	m/si	3
man standing	ǀ	m/stg	5
female prone	⌐●	f/pr	2
female sitting or squatting	⌐	f/si	4
female standing	ǀ	f/stdg	6

Once memorized, the number code is the easiest and quickest to use. Throughout the system, the active subject is indicated first. A man talking to a woman while both are standing is recorded as 56, while 65 means the woman is talking to the man. When it is not possible to tell which subject is more active, parentheses are used. Whenever there are extremes in age, size, or status, these should be noted.

The Sociofugal-Sociopetal Axis (SFP axis)

Osmond (1957) uses the terms *sociofugal* and *sociopetal* to describe spatial arrangements or orientations that push people apart and pull them in—orientations that separate and combine people, that increase interaction or decrease it.

As can be seen, the SFP axis is a function of the relations of the bodies to each other. In theory, there are endless variations in the orientation of two bodies in relation to each other. However, in proxemic transcription the observer is interested in recording only those distinctions which are operationally relevant to the participant. After experimenting unsuccessfully with a number of overly sensitive systems, an 8-point compass face was finally selected as the most appropriate model (see figure 1).

Zero and 8 are placed at North, 2 at East, 4 at South, 6 at West. Zero represents two subjects face to face (maximum sociopetality), 8 two subjects back to back (sociofugal), 2 two subjects slightly facing but at right angles so they can either see each other peripherally if they look straight ahead or look in each others' eyes. Position 4, in which the subjects stand side by side with the north-south axis running through a parallel to their shoulders, is also very common in the United States. Orientation 6 is definitely sociofugal because the subjects' shoulders are at right angles but with the faces pointing out and away; in order to see each other, they must crane their necks.

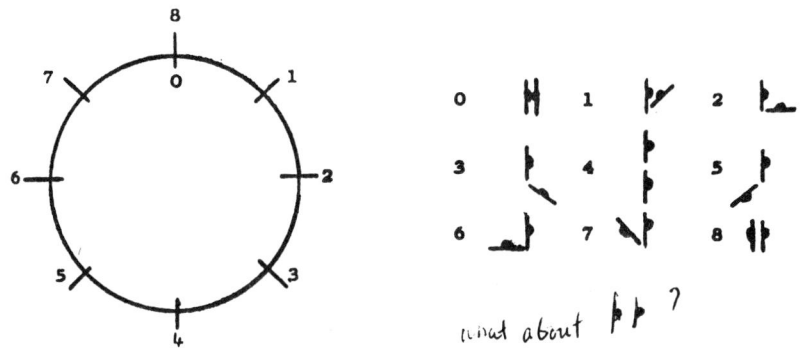

FIG. 1. SFP axis notation code.

Which components of the SFP axis are favored and for what transactions, is largely culturally determined. These components are also linked with the social setting and the age, status, and sex of the two parties. On the basis of continuous observation over the past three years, it is possible to offer some generalizations concerning the principal structure points of the American system. In interpreting these generalizations the reader should keep in mind that, like other communications, proxemic communications are always read in context and have no meaning independent of context.

0, 1, 2, 4 and 8 are most frequently observed. 0 is for direct communications where the intent of one or both of the participants is to reach the other with maximum intensity.

2 is more casual and less involved. A subject of common interest is often discussed using this axis. The subjects may shift to 0 or 4, depending on how involved or uninvolved they become.

4—the shoulder-to-shoulder axis—is one in which two people are normally watching and/or discussing something outside themselves, such as an athletic event or the girls going by on a Saturday afternoon, without necessarily being involved with each other. This is the axis for very informal, transitory communications.

Position 6 is used as a means of disengaging oneself. It is not quite, but almost as, sociofugal as 8.

The Kinesthetic Factors

One of the most basic forms of relating in space, one which is deeply imbedded in man's philogenetic past, is the potential to strike, hold, caress, or groom. In threatening situations among animals, enemies and potential enemies are not permitted within striking distance (Hediger 1955).

This applies in intra-species as well as interpersonal relations. A cowboy walking around a horse illustrates this principle; he uses three different distances. With a strange horse he follows an arc just outside the radius of the horse's hoofs when kicking. With familiar horses—those not known to be dangerous—he walks inside this circle but not too close. With his own horse he uses a much closer (more intimate) distance and may even brush the horse's tail as he passes.

The kinesthetic code and notation system is based on what people can do with their arms, legs, and bodies, and the memory of past experiences with one's own as well as other bodies. Another person is perceived as close (as we shall see later) not only because one may be able in some instances to even feel the heat radiating from him, but also because there is the potential for holding, caressing, or of being struck.

Basically there are four ways of relating with the body.[19] These are a function of four basic inventories of potential actions: 1) touching with the head or trunk; 2) touching with forearms, elbows, or knees; 3) touching with arm fully extended; and 4) with the arm and the leg extended and body leaning, *i.e.*, stretching (far apart but still able to touch). These are noted and/or symbolized as follows:

Each increment symbolizes a progressively greater distance. The *measured* distance depends largely on the size and shape of the individuals involved and the SFP axis, so that the figures given below are approximations, empirically derived from a small sample of medium-sized persons. The important feature is that, to the persons involved, the distance is perceived in terms of the capabilities of the two bodies. The observer therefore will have to revise the figures upward or downward depending on his own body build, particularly in relation to the larger distances.

#1 0"–3"   #2 15"±2"   #3 22"±3"   #4 40"±4"

It is possible to code kinesthetic relationships in two ways: A) As one of the four distances, or B) as one of the four distances plus some space. This provides an inventory of eight distances which can be recorded in the following manner using a number code:

- # 1 within body contact distance
- #10 just outside body contact distance
- # 2 within easy touching distance with only forearm extended
- #20 just outside forearm distance ("elbow room")
- # 3 within touching or grasping distance with the arms fully extended
- #30 just outside this distance
- # 4 within reaching distance
- #40 just outside reaching distance

Since two parties are involved and each has his own repertoire of the 8 kinesthetic distances listed, it is possible to construct a matrix with 8 dimensions on a side (one for each kinesthetic distance) which contains 64 different slots (8×8). Because such a matrix is nothing more than a mechanical way of insuring that all possible combinations have been accounted for, there is considerable duplication (13 and 31 for example). From these 64 combinations 11 basic distinctions have proved sufficient to account for all the space transactions observed to date. These are given below in figure 2.

Symbols	code #
∥	11
∥	101
⋈	12
⋈	102
⋈	22
⊓	13
⊓	103
⊓	33
⊓	303
⋀	44
⋀	404
... outside the system when extensions are introduced, such as swords, bolos, blow guns, and modern arms.	55

FIG. 2. Kinesthetic code and notation system.*

Touching

Cultures vary greatly in the amount of touching which occurs between people. Even in the United States, there are groups which participate in considerable touching and others whose members assiduously avoid touching anyone but those with whom they are intimate.

A seven-point scale seems sufficient for the moment to code the majority of contact-non-contact situations. Since it is possible for each person to touch the other, all combinations can be recorded on a 7×7 grid.

This proceeds from 00 mutual caressing to 66 in which there is no contact by either party (see figure 3).

---

* To speed recording it is recommended that the number code be committed to memory, which should require only a matter of minutes.

Vision

The role of vision in judging distance and in communication is incredibly complex. The non-specialist looking at the field as an outsider finds conflicting statements concerning not only the eye itself but the entire process of vision. There is agreement, however, that the visual sense is the most complex and highly evolved of the senses. Depending on the source one chooses, and using the size of the channel of the brain as a rough index of capacity, the eye feeds from 6 to 20 times as much information to the brain as the ear (Gibson 1950; LeGrand 1957). How the eye is used as a function of one's culture is regulated formally, informally, and technically (Hall 1959). That is, the culture specifies at what, at whom, and how one looks, as well as the amount of communication that takes place via the eye.

For example, a Navaho is taught to avoid gazing directly at others during conversations. He also avoids looking at his mother-in-law, but distinguishes between these two events. The Greeks, on the other hand, emphasize the use of the eyes and look for answers in each other's eyes (their intent gaze can be disturbing to the American). Americans often convey the impression to the Arabs that they are ashamed. The way Americans look at the other person during conversations is the principal reason given by Arabs for this impression.

Given the above distinctions and the fact that culture is learned, the anthropologist must pose the following question: Is it possible, by some simple means, to distinguish operationally between these different events? Part of the answer lies in the structure of the retina. The fact that the retina provides for three distinct and easily identifiable types of vision makes it possible to make distinctions of the type indicated, and at the same time provides us with the basis for a notation system.

The Three Areas of the Retina

In the middle of the visual field there is a small pit in the retina called the fovea. It subtends a visual angle of only 1°. Some idea of the fineness of visual detail made possible by this structure can be gained by considering the fact

	0	1	2	3	4	5	6
0	00	01	02	03	04	05	06
1	10	11	12	13	14	15	16
2	20	21	22	23	24	25	26
3	30	31	32	33	34	35	36
4	40	41	42	43	44	45	46
5	50	51	52	53	54	55	56
6	60	61	62	63	64	65	66

0 = caressing and holding
1 = feeling or caressing
2 = extended or prolonged holding
3 = holding
4 = spot touching (hand peck)
5 = accidental touching (brushing)
6 = no contact whatever

FIG. 3. Touch code.

	f	m	p	0
f	ff	fm	fp	f0
m	ms	mm	mp	m0
p	pf	pm	pp	p0
0	0f	0m	0p	00

	1	2	3	8
1	11	12	13	18
2	21	22	23	28
3	31	32	33	38
8	81	82	83	88

FIG. 4. Visual code.

that there are 25,000 closely packed cones, each of which is connected to a single bipolar nerve cell. (There are no rods in the fovea.)

Surrounding the fovea is an oval area called the macula, with a vertical visual angle of approximately 3° and a horizontal visual angle of approximately 12°. This is the area of clear vision. From the macula on out, vision becomes less and less distinct. According to Dr. Milton Whitcomb, secretary to the National Research Council's Committee on Vision, approximately half of 130,000,000 rods and all the 7,000,000 cones are concentrated in a central portion of the retina covering a visual angle of 20°. Peripherally it is possible for most people to perceive motion laterally on the temporal side of each eye at 180° or better.

The eye therefore provides three different ways of viewing, depending on where the image falls on the retina: the fovea, the macula, or the periphery. A fourth alternative is to screen out vision.

Therefore it is possible to code these as:

f = foveal (sharp)   1
m = macular (clear)   2
p = peripheral   3
0 = no visual contact   8

A 4×4 grid illustrates the 16 combinations of the ways that people look at each other (see figure 4).

In constructing the visual scale and notation system it was deemed advisable to depart from previous practice and allow space for later additional notations. Therefore the numbers 0, 4, 5, 6 and 7 were reserved for this purpose.

To return to our earlier examples, two Navahos talking would be recorded as 33 (pp). A Navaho and his mother-in-law as 88 or 38 (38 because, while neither is supposed to look at the other, the mother-in-law is apparently dominant, since others will tell the son-in-law when she is around so he can leave or turn his back).[20]

Both Arabs and Greeks tend to read the other person's eye (11) much more than Americans.

In making a record of the way that people look at each other, it will be necessary for the field worker to develop recording techniques consistent with visual interaction patterns of the culture under observation. The capacity of virtually any subject to determine where another person is looking is extraordinarily well developed. According to Gibson the "gaze line" can be calculated with accuracy approaching sensory acuity.[21]

The field worker can therefore rely on his own ability to determine where subjects are looking. The point is, however, that the topic of looks and of looking must be completely explored with one's subjects and backed up by observations. Practice will improve one's capacity to record visual interactions. The technique used by this field worker is to try to catch the most characteristic sequence of looks for a given type of transaction recording three or four, one above the other pp

mm	22
pp	33.
mm	22

Virtually nothing is known of vision as a factor in human transactions. The most critical need is for data.

The Thermal Factors

Although not much is known about the thermal zone or zones, heat gain and loss apparently influence the structuring of close distances.[22] As indicated elsewhere, responses to heat flow in the subtle sense are not usually considered in a distance-regulating context. Operating as it does, almost totally out-of-awareness, heat flow is traditionally thought of as strictly a matter of comfort. Nevertheless it has been observed and commented on by some subjects that the sensing of heat from another body can result in a movement either towards or away from the source.

The degree to which this type of response occurs out-of-awareness is illustrated in the recording of a trivial event that might well have otherwise gone unnoticed. The ambient situation was a dinner party. There was the usual semi-crowding with the attention on conversation. This observer, hand resting on table, was suddenly aware of a rapid reflex-withdrawal of his hand. There was no immediate sensation such as pain or touch. Visual examination and quick review of the preceding few seconds revealed two things: 1) the hand of another guest approximately $2\frac{1}{2}$ inches (6.5 cm) from where his own hand had been a moment before, and 2) a memory trace of having detected heat in the hand. A replacing of the observer's hand in the same area revealed that heat flowing from the other person's still stationary hand was definitely detectable.

It is possible to code heat flow (digitally) as either being detected or not, depending on whether the two parties are within each other's thermal sphere or not. Heat flow is detected in two ways, by radiation and by conduction.

The thermal code and notation system is as follows:

conducted heat detected	thc	1
radiant heat detected	thr	2
heat probably detected	th	3
heat not detected	t̸h̸	8

Because so little is known about thermal responsiveness to others, numbers

0, 4, 5, 6, and 7 are reserved for future use. The field worker will have to experiment until he has subjects who can tell him when they have detected heat. As with the level of loudness of the voice, there is considerable range in heat output. Clothing of course reduces sensitivity somewhat. One is dealing, however, with a primitive communication system that has been overlooked by social scientists but may have common currency. A limited number of subjects, for example, stated they were fully aware of the significance of changes in body temperature in others which could be interpreted as readily as though words had been used. One example will illustrate this point. The subject, a rather voluptuous young female, noted that when she danced with certain young males she could detect a rise in temperature in the abdominal region which foreshadowed genital tumescence and which could be differentiated from a temperature rise due to exertion. What is more, this rise in temperature could be detected several inches from the dance partner.

It should be noted that the body does not heat uniformly, but in specific areas depending on the situation. The whole subject of mapping the geography of heat output of the body under different emotional states remains to be explored.

Olfaction

In the United States advertisements extolling the virtues of not offending by the odor of one's body or breath are a prominent feature of American mass media. That is, the olfactory sense is culturally suppressed to a greater degree than any of the other senses.

Pleasant odors, such as perfume on women or bay rum on men, are desirable but should not be detectable at more than intimate distance for the middle class. In fact, when olfaction is present it usually signals intimacy.

Sexual odors, bad-smelling feet, and flatus are definitely taboo in all but a very limited number of situations and relationships. The degree to which American culture has dulled, repressed, or dissociated the olfactory capacities is not known; much more comparative data is needed.

Also, as with vision and hearing, there is a great range of olfactory acuity. It is highly probable that sensitivity to different odors may be selective in a culturally patterned way, so that what is quite obvious or even overwhelming to a foreigner may not be *significant* to a local subject. Because of our own taboos, American research workers will have to be particularly careful to check their own observations with local colleagues.

In light of all this, and given the meager state of our present knowledge of olfaction as a proxemic indicator, it was decided to simplify the recording procedure as much as possible.

At any given distance, only four observations and accompanying notations are made:

OLFACTORY CODE

differentiated body odor detectable	dbo or 1
undifferentiated body odor detectable	ubo or 2

breath detectable	br	or 3
olfaction probably present	oo	or 4
olfaction not present	∅	or 8

In essence, the investigator looks for boundaries and whether they have been crossed or not. Everyone is surrounded by a small cloud or haze of smell, varying in size according to physical setting, emotional state, and culturally prescribed norms. The investigator must determine at what point the smell is unmistakable and where this fits into the total proxemic picture. Usually there is little ambiguity. Most transactions occur either inside or outside these boundaries.

In recording breath, it must be determined at what point people feel free about directing their breath and whether the warmth and moisture of the breath is sought after or not.

Use code designation 4 (oo) when it is pretty certain that olfaction is present but otherwise unspecified. The numbers 0, 5, 6, and 7 are reserved for future refinements.

Questioning subjects on olfaction has not proved a problem to date. All Arabs interviewed spoke freely about bad breath and feet odors and how these must be avoided. Friends and relatives tell them when they should not stand too close to people. Normally they do not feel close to people until they can detect the heat, moisture and smell of breath. There seems to be little doubt that the Arab employs olfactory cues to set distance. The principal difference between the Arab and the American patterns is that for Americans to be within smelling distance is to introduce intimacy, whereas with the Arab it apparently only makes them feel "at home." Without smell, Arabs apparently feel somewhat "left out." It should be noted, however, that the Arab data is based on such a small segment of the over-all culture that any interpretations made at this point must be tentative.

Voice Loudness

The loudness of the voice is modified to conform to culturally prescribed norms for a) distance, b) relationship between the parties involved, and c) the situation or subject being discussed. Cultural norms can vary for any one of the three as well as for all three. The same applies to sub-cultures within a larger cultural frame. Voice level, therefore, is relevant as a significant variable in judging distance.

Holding the other two dependent variables—relationship and situation—neutral and constant, an American will normally code a whisper as close and a shout as distant. Similarly, Brodey's blind subjects judge the distance of a speaker by the loudness of his voice.[23] This is, of course, not the only means by which the blind judge distance; it is, however, an important one.

The culture of one's upbringing has a good deal to do with how loudness is perceived. For example, as a general rule Arabs sound loud to Americans. Arabs, on the other hand, will comment among themselves that the American's voice is too low and sounds insincere. Subjects' unguarded comments on ethnically associated loudness of voice are not the only source of data on this subject. Children have to be systematically taught not only what is correct and incorrect usage but how to modulate properly the loudness of the voice.

The investigator can provide a good deal of clinical data from his own past. No standards have been established for judging voice loudness *except* those people learn and against which they judge the behavior of others. There is no alternative except to code the loudness of voice employing the investigator's own culturally calibrated measuring device.

Using the investigator as a measuring device may not satisfy the rigid requirements of all scientists. It should be kept in mind, however, that this is what all people do whenever they respond to loudness or softness of the voice. The principle of measurement is the same as two people standing back to back to see who is the taller. The field investigator should test his own evaluations against those of others. This can be accomplished easily by having two or more observers record the same transaction separately and comparing notes.

Seven degrees of loudness have proved sufficient to code all vocal transactions to date. (A zero for silence increases the number from six to seven.)

VOICE LOUDNESS SCALE

descriptive level	mnemonic code	number code
silent	0	0
very soft	vs	1
soft	s	2
normal	n	3
normal+	n+	4
loud	l	5
very loud	vl	6

Cautions and Reminders

It is important not to overcomplicate the recording of voice loudness. As with other features of vocalization, over-all loudness varies with the individual.

Most people are aware, however, when the speakers of another language in any given setting sound louder or softer than the speech they are used to hearing at home or when a person of their own culture speaks overly slowly or softly, as this may signify anger. One of the best ways of bringing home the point that voice level conforms to cultural norms is to be around small boys at the dinner table before they have learned to modulate their voices. While this task seems endless to parents, the seal of culture is well impressed by age 11 to 14 for most normal situations.

In the field one should ask one's subjects for an appraisal of voice loudness. "Are those people—or is he—or she talking in a normal tone of voice?" This will usually elicit a comment if there is anything unusual about the level of voice and provides added information. It is possible, for example, that variation of voice loudness is not as important as it is in the United States, or is, perhaps, more important. One English subject interviewed over a long period of time turned out to have a remarkable capacity to modulate and direct his voice in such a way that it was difficult to tell how far away he was.

Language Style

Traditionally any layman will affirm that what one talks about and the manner of talking are linked with distance and situation, but he will be unable to describe what the differences are.

The linguist Joos (1962) has provided Americans with an analysis of their own linguistic behavior as viewed through a situational screen. The degree to which other cultures recognize and talk about situational styles or dialects is not known.

Joos lists five styles, each used for a different situation. They are: intimate, casual, consultative, formal, and frozen, and while the matching with distance zones seems close but not perfect, more needs to be known about the proxemic aspects of style.

If different styles of speech are recognized and used in specific situations (classical and colloquial Arabic, for example), these should be noted also by the observer of proxemic behavior.

PROXEMIC SYSTEMS AS COMMUNICATION

Hockett (1958) defines communication as any event that triggers another organism. While many other life forms communicate (as for example when a bee informs another where honey is, by means of an orientating series of dance steps), language is characteristically human. Hockett lists seven principal features of language: duality, productivity, arbitrariness, interchangeability, specialization, and displacement, and cultural (not genetic) transmission.

Proxemic behavior is obviously *not* language and will not do what language will do. Nevertheless a careful analysis demonstrates that proxemic communication as a culturally elaborated system incorporates more features named by Hockett than one might suppose. For example, language is both "plerematic" and "cenematic"—i.e. has both sets and isolates (Hall 1959) or units that build up, or combine, to form a different kind of unit.

Proxemics lacks none of the seven features of language listed by Hockett. Its arbitrariness is not obvious at first, because proxemic behavior tends to be experienced as iconic—e.g., a feeling of "closeness" is often accompanied by physical closeness—yet it is the very arbitrariness of man's behavior in space that throws him off when he tries to interpret the behavior of others across cultural lines. For example, the fact that Europeans name streets (the lines that connect points) and the Japanese name the points and ignore the lines, is arbitrary. The fact that in the European cultures people arrange objects whereas in Japan they arrange spaces, is arbitrary. American suppression and repression of olfaction in proxemic behavior is also arbitrary. The arbitrariness operates on a Whorfian level rather than a more conscious level. Hence it is even more difficult to come to grips with than lexical items.

Proxemics demonstrates duality of a primitive but nevertheless readily identifiable sort. The units (cenemes) build up. For example, the elimination or introduction of visual contact can completely redefine physical closeness. The operating principle behind the confessional booth is that it makes it possible to bring a man and woman together in an intimate setting but without the ability to touch or see each other. Removing the partitions in the confessional would completely redefine the situation, particularly if it were an enclosed confessional.

Interchangeability means that subject "A" can play "B's" part and vice versa. In other words, the subject and the communication are not irrevocably tied together, as is the case with the male peacock and his display. A feature of proxemic communication is its interchangeability.

Displacement refers to the capacity of language to deal with displacements in time or space. In the animal world territorial markers (particularly the olfactory ones) characteristically feature displacement. In man fixed and semi-fixed features also feature displacement (see Hall 1963). Boundary markers, fences, closed doors, chairs placed in a conversational group, the arrangement of furniture in an auditorium, the psychiatrist's couch, and the layout of offices all enable someone *who knows the system* to interpret what has taken place or the message that is intended. As Hazard notes: "Walk into an empty courtroom and look around. The furniture arrangement will tell you at a glance who has what authority" (Hazard 1962). Only dynamic space (the actual distances between people when they are interacting such as *tone of voice*) lacks the displacement features.

Specialization refers to the fact that language tends to refer to specific items or events, *i.e.*, to become "specialized." Proxemic behavior is *not* as highly specialized as language. Nevertheless it contains great capacities for specialization. The American pattern for comforting and lovemaking are seldom confused, even though both involve great closeness. The fact that "duality" is present, that there are differences in the interplay of receptors in these two instances (avoidance of olfaction in comforting for one thing) makes incipient specialization possible. Nothing could be more specialized than the sacredness or taboos associated by all people with certain specified places like Mecca, the Navaho mountain, or the Chindi hogan. What is more specialized than a boundary, or father's chair, the "head" of the table, the tokonoma in the Japanese house, the proper distance to be maintained when attracting attention of someone without intruding, or the distinction between the relative and non-relative side of the office in the Middle-East (Hall 1959: ch. X)?

In other words, proxemic behavior parallels language, feature for feature. It is, however, much *less* specialized and more iconic. It tends to be treated as though certain features associated with language were lacking. The iconic features of proxemics are exaggerated in the minds of those who have not had extensive and deep cross-cultural experience. In fact, when a subject stops treating proxemic behavior as iconic and sees its arbitrariness, he is beginning to experience the over-all arbitrariness of culture.

Sebeok (1962) presents the hypothesis that animal communication is most

1) *Postural—sex identifier*

male 1, 3, 5

female 2, 4, 6

2) *Orientation of bodies* (SFP axis)

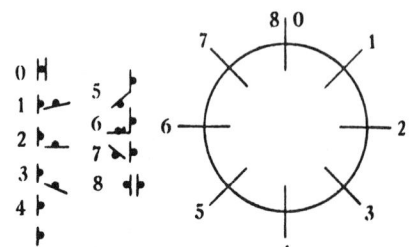

0, 1, 2, 3, 4, 5, 6, 7, 8

3) *Kinesthetic factors*

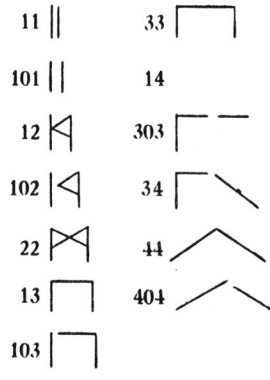

11, 101, 12, 102, 22, 13, 103

33, 14, 30.3, 34, 44, 404

4) *Touch code*

caressing & holding	0
feeling or caressing	1
prolongéd holding	2
holding or pressing against	3
spot touching	4
accidental brushing	5
no contact	6

5) *Retinal combinations* (Visual code)

foveal	f	1
macular (clear)	m	2
peripheral	p	3
no contact	nc	8

6) *Thermal code*

contact heat	thc	1
radiant heat	thr	2
probable heat	th	3
no heat	th̷	8

7) *Olfaction code*

differentiated body odors detectable	do	1
undifferentiated body odors detectable	ubo	2
breath detectable	br	3
olfaction probably present	oo	4
olfaction not present	∅	8

8) *Voice loudness scale*

silence	si	0
very soft	vs	1
soft	s	2
normal	n	3
normal↗	n↗	4
loud	l	5
very loud	vl	6

FIG. 5. Key to combined proxemic notation system.

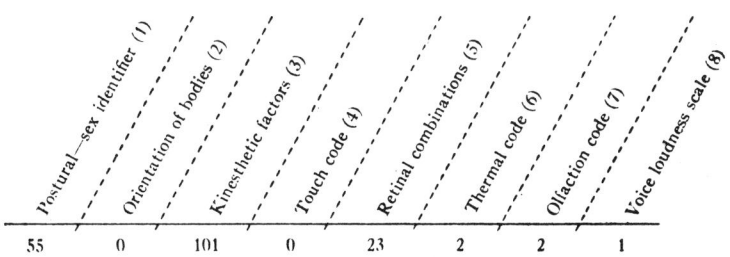

1) Two men standing;
2) facing each other directly;
3) close enough so that hands can reach almost any part of the trunk;
4) touch does not play any part;
5) man speaking looking at, but not in the eye, partner only viewing speaker peripherally;
6) close enough so that radiant heat would have been detected;
7) body odor but not breath detectable;
8) voice very soft.

FIG. 6. Recorded transaction and key.

often "coded analogly," whereas speech is coded both digitally and analogly. Proxemic behavior is also coded both digitally and analogly.

The knowledge of the relationship of language to other cultural systems—not to mention the communications systems of lower life forms—is in a state of flux. Better understanding of how these other systems function and how best to study and describe them, in part awaits the development of improved procedures for discourse analysis.

Language is most commonly treated as an instrument for communicating *from* one person *to* another, rather than as a transaction; Hockett (1958), Hymes (1962) and Sebeok (1962) are exceptions though for different reasons. Joos (1962) comes closest to a transactional view with his five linguistic styles of English. Proxemic behavior, on the other hand, by its very nature inevitably is reduced to a transaction—a transaction between two or more parties, or one or more parties, and the environment. It is this very feature that makes it difficult to relate proxemics at all significant levels to current linguistic models.

Concerning the relationship of the *etic* level of analysis to the *emic* level and how one proceeds from the former to the latter (Hymes 1962), it will be apparent to the reader that this presentation is concerned more with the proxetics than proxemics, and is therefore only the first of a series of steps in a long complex process.

In the course of this investigation attempts to classify behavior on the emic level were not successful until a system of observations and notations had been developed that enabled the observer to account for two types of differences: A, between events in contrasting systems, and B, between one proxeme and another within a given system. Thus one of the points of contrast observed by

Americans overseas when interacting with a variety of peoples in the Mediterranean culture areas, is in the direction of the breath during personal (but not intimate) conversations. Americans are taught not to breath on people, particularly strangers or people of higher status, and are made aware of this when others breathe on them. (Attention to the structure points of the system is also drawn during acculturation of the young, so that the anthropologist can learn a good deal about the American system while correcting his children.)

Within the bounds of the American system there is considerable variation in the use of the eyes and the degree of touching that is permissible by strangers. Touching or looking of a type that falls outside one's familiar pattern may be coded as unusual in the same manner that we treat unfamiliar alophones of a given phoneme and still be recognized as a permissible variant of a familiar form. In one encounter, for example, items 5 (accidental brushing) and 6 (no contact) of the touch code (scale #4), were aloproxes of the same proxeme for the writer, but were coded as different proxemes by an upper middle class Dutch subject.

SUMMARY

The notation system presented here is designed to provide a way of being rather specific when talking about observations of a very limited nature. No claim is made for the superiority of this system. It has proved to be reasonably workable and simple. If it persists at all as a tool in the hands of the ethnographer, it will undoubtedly go through transformations as use reveals its defects. Currently the visual dimension stands out, out of all those described, as the one most requiring additional treatment.

The following visual aids have been prepared to summarize the data of this article:

Figure 5 shows the entire code with all its component parts in the order in which they should be recorded.

Figure 6 shows a sample of a record of a transaction.

NOTES

[1] Research supported by grants from the National Institute of Mental Health and the Wenner-Gren Foundation for Anthropological Research.

[2] When this research was conceived, no suitable designation had been found for the study of microspace as a system of bio-communication. *Human topology*, *chaology* (the study of boundaries), *choriology* (the study of organized space), and others were considered. *Proxemics* was chosen because it suggests the subject to the reader.

[3] In an inter-cultural encounter the structural details of the two culture systems combine in one of three ways: A) They can mesh or complement, so that the transaction (Kilpatrick 1961) continues or is reinforced. B) They can clash or interfere which has an inhibiting effect on the transaction. C) They can be unrelated so they neither reinforce nor inhibit the transaction.

Even within the broader context of American culture, it is possible to observe these three types of interaction; for example, the Utah Mormon's version of time—in which there is virtually no leaway for being late—meshes nicely with the U. S. military system.

Interference can be observed whenever there is an attempt to integrate two groups of individuals, one having internalized the diffuse point pattern while the other uses the displaced point pattern (Hall 1959).

As a general rule, the time systems of middle class men and women from the same subculture tend to operate in such a way that there is a minimum of interference, even though the two systems differ. During the day, and for business, male time takes precedence. In the late afternoon and evening—particularly for meals and social occasions—women's time is the dominant pattern. Cultural interference is analogous to linguistic interference as described by Weinreich (1953).

[4] Reducing distortion and minimizing contamination of data has long been considered a basic feature of the methodology of the physical sciences. Possibly because of the great complexity of our data, most anthropologists until recently have avoided the issue of distortion. Distortion, incidentally, should not be confused with accuracy of reporting, or lack of accuracy. Reporting is something the anthropologist does. Distortions are in the hands of the informant.

One might argue that the distortions on the overt level are distortions of content only and do not hide the patterns, that anything an informant tells the anthropologist is grist for the anthropologist's mill, or that the distortions themselves provide insights into the culture—all of which is often true.

Nevertheless, if one is seeking to construct a theory of the culture one is studying (Goodenough 1956), this process proceeds more rapidly if the building blocks used to construct that theory are reasonably stable. It should be noted that the stories made up by an informant are of necessity rooted in his experience and as such are representations of his culture. However, the *design* he creates is of short duration, i.e. tomorrow he will tell a slightly different story. The basic distinction between out-of-awareness patterns that cannot be cut to conform to situational demands and the conscious screening of the truth from an outsider is in the time that two types of events remain stable. In addition to the desirability of increasing stability of the components that go to make up one's theory of a given culture, there is the practical matter of building confidence in one's subjects. In most instances ability to control the communications systems marks one as an insider; absence of this control marks one as an outsider. Any field anthropologist who has experienced the pride his subjects show in his own increased knowledge of, and skill in, controlling their culture, knows the importance of being able to use a given system correctly. Reduction of distortions and levels of awareness constitute the topic of another publication. The subject is mentioned here only to indicate that the writer is not unaware of the complexities and as a reminder that the most mundane and taken-for-granted assumptions often turn out to be most difficult to come to grips with.

[5] Goffman (1959) describes still another level of awareness that deals with the mask one wears in order to play the proper parts in daily transactions.

[6] Hymes, in commenting on this paper (also see Hymes 1962), suggests how some "functionally relevant dimensions" can be identified from the discomfiture of the subject when patterns have been broken.

[7] Individual and regional diversity in proxemic patterns is comparable to that encountered in the use of time, materials, and language. Distinctions on these levels are not relevant to this presentation. Instead, a more basic pattern should be mentioned: Americans of European ancestry fall generally into two groups—contact and non-contact. Non-contact Americans minimize physical contact—touch or holding during encounters when the transaction is social or consultative in nature. Contact Americans, on the other hand, employ touching and holding which is sufficiently different from the former pattern as to cause comments. Hereafter, whenever the term "American" is used, it refers only to the dominant non-contact group.

[8] Thiel (1961) has recently developed and published a system for describing the kind of space architects and landscape architects deal with. See also Goldfinger 1941 and 1942.

[9] Benedict (1946) describes how the anthropologist not only uses informants but draws upon every other available expression of the culture, including art forms, movies, and literature. It is in the tradition of anthropology, therefore, that the anthropologists look to other fields, particularly art and literature, as a means of checking their own observations.

[10] See Gibson 1950. This process, however, seems to be different from the constant process of adjustment to another person who does not stand still but moves. Little is known about the former, even less of the latter.

[11] Concerning the relationship between the encounter and the setting, or between fixed feature

space and dynamic space (Hall 1963), there seems little doubt that such a relationship exists. The evidence of the animal psychologist and the ethologist is firmer than for man. However, on the human level, data from widely scattered sources points in the same direction (Hall Mss). Until now, man's behavior in space has been treated from a strictly physical-anatomical point of view, and with the implicit assumption that cultural differences did not exist, and that if they did they were unimportant.

[12] Specifically, the work of linguists in the tradition of Edward Sapir, Leonard Bloomfield, and Benjamin Lee Whorf.

[13] My contact with the various psychoanalytic movements has been eclectic. The out-of-awareness features of communication and the use of one's self as a control are so important that it is difficult to credit them sufficiently. Both are pivotal in any research this investigator undertakes (See Fromm 1941 and Sullivan 1940).

[14] Frake, reviewing Goodenough's thesis (1962), sets forth many of the conditions which must be met in the writing of "productive ethnographies" and the absolute necessity of tapping the *cognitive world of one's informants*, and avoidance of "a priori notions of pertinent descriptive categories." (italics mine)

[15] The proxeme equates with the phoneme of language, but on a much lower and simpler organizational level (cf. Hall 1959, ch. 6-8).

[16] Linguistic style (Joos 1962) associated with each distance represents the most recent and least known and also apparently the most complex of the subsystems linked to proxemic patterns. At present, Joos' type of analysis is only available for English. It is mentioned here as a reminder to the field worker, since other languages may have been subjected to similar analysis.

[17] The information is considerably less than that transmitted for one weather station code used to plot the basic data for weather maps. These codes can be transcribed on a weather map as fast as a man can enscribe the simple weather symbols.

[18] Figure given in response to question at Interdisciplinary Work Conference on Paralanguage and Kinesics, University of Indiana 1962.

[19] Given the great flexibility of the body it may seem strange that the number of relational possibilities is so small. The situation is comparable to that noted earlier under sex and posture. Four is sufficient to record all distinctive features in the kinesthetic inventory. Any more would be too complex, as each of the four is coded digitally (touching or not touching) which yields 64 possible combinations, only 11 of which are really essential.

[20] The ethnographic data was obtained during field work with the Navaho some 30 years ago in the Pinyon Black Mesa region of the reservation. It is possible that the details of the mother-in-law taboo have changed since that time.

[21] Gibson, J. J., letter to the investigator dated September 5, 1962.

[22] Recent studies in thermography (the study of infra-red radiation) demonstrate that human skin is an ideal emitter and receiver of infra-red energy (Barnes 1963). How effective it is can be seen by looking at a scale of emissivities of various substance in which emissivity values have been given:

mirrors and polished metals	.02–.03
polished lead and cast iron	.21–.28
black loam and fire brick	.66–.69
wool and lumber	.78
lamp black and soot	.95
human skin	0.99

[23] The perceptual world of five blind subjects has been investigated systematically for the past 2½ years by Dr. Warren Brodey of the Washington School of Psychiatry. This investigator has participated in the research.

REFERENCES CITED

BARKER, R. G. AND BARKER, L. S.
    1961 Behavior units for the comparative study of cultures. *In* Studying personality cross-culturally, ed. by B. Kaplan. New York, Row Peterson.

BARKER, R. G. AND WRIGHT, H. F.
    1954 Midwest and its children; the psychological ecology of an American town. Evanston, Row Peterson.

BARNES, R. B.
    1963. Thermography of the human body. Science 140:3569:870–877, May 24.

BENEDICT, R.
    1946 The chrysanthemum and the sword. Boston, Houghton Mifflin.

BIRDWHISTELL, R.
    1952 Introduction to kinesics. Louisville, University of Louisville Press.

BOAS, F.
    1911 Introduction, Handbook of American Indian languages. B.A.E. Bull. 40. Washington, D. C., Smithsonian Institution.

DORNER, A.
    1958 The way beyond art. New York, New York University Press.

FRAKE, C.
    1960 Family and kinship among the eastern Subanun. *In* Social structure in Southeast Asia, G. P. Murdock, ed. New York, Viking Fund Publications in Anthropology No. 29.
    1962 Cultural ecology and ethnography. American Anthropologist 64:53–59.

FROMM, E.
    1941 Escape from freedom. New York, Rinehart.

GIBSON, J. J.
    1950 The perception of the visual world. Cambridge, Harvard University Press.

GOFFMAN, E.
    1957 Alienation from interaction. Human Relations 10:1:47–60.
    1959 The presentation of self in everyday life. New York, Doubleday.

GOLDFINGER, E.
    1941 The sensation of space. Urbanism and spatial order. Architectural Review, November:129–131.
    1942 The elements of enclosed space. Architectural Review, January:5–9.

GOODENOUGH, W. H.
    1956 Residence rules. Southwestern Journal of Anthropology 12:22–37.
    1957 Seventh annual round table meeting on linguistics and language study, Garvin, ed. Washington, Institute of Languages and Linguistics, Georgetown University. Monograph series on languages and linguistics no. 9:167–177.

GROSSER, M.
    1951 The painter's eye. New York, Rinehart.

HALL, E. T.
    1955 The anthropology of manners. Scientific American 192:85–89.
    1959 The silent language. New York, Doubleday.
    1963 Proxemics—the study of man's spatial relations. *In* Man's image in medicine and anthropology. International Universities Press. New York.
    1964 Spatial features of man's biotope. Mss to appear *in* Proceedings of the Association for Research in Nervous and Mental Diseases. New York.

HALL, E. T. AND TRAGER, G.
    1953 The analysis of culture. Washington, American Council of Learned Societies.

HAZARD, J.
    1962 Furniture arrangement and judicial roles. ETC 19:181–188.

HEDIGER, H.
    1955 Studies of the psychology and behaviour of captive animals in zoos and circuses. London, Butterworths Scientific Publications.
    1961 The evolution of territorial behavior. *In* Social life of early man, ed. by S. L. Washburn. New York, Viking Fund Publications in Anthropology No. 31.

HEWES, G.
    1955 World distribution of certain postural habits. American Anthropologist 57:231–234.

HOCKETT, C.
    1958   A course in modern linguistics. New York, Macmillan.

HYMES, D.
    1962   The ethnography of speaking. *In* Anthropology and human behavior, ed. by Gladwin and Sturtevant. Washington, D. C., The anthropological Society of Washington.

JOOS, M.
    1962   The five clocks. International Journal of American Linguistics 28:pt. 5.

KILPATRICK, F., ed.
    1961   Explorations in transactional psychology. New York, New York University Press.

LE GRAND, Y.
    1957   Light, color, and vision. London, Chapman and Hall Ltd.

LYNCH, K.
    1960   The image of the city. Cambridge, Technology Press and Harvard University Press.

McQUOWN, H.
    1957   Linguistic transcription and specification of psychiatric interview materials. Psychiatry 20:79–86.

OSMOND, H.
    1957   Function as the basis of psychiatric ward design. Mental Hospitals:23–29.

SEBEOK, T.
    1962   Evolution of signalling behavior. Behavioral Science 7:430–442.

SULLIVAN, H. S.
    1940 and 1945   Conceptions of modern psychiatry. Washington, Wiliam Alanson White Foundation.

THIEL, P.
    1961   A sequence-experience notation for architectural and urban space. Town planning review, April:33–52.

WALLACE, A.
    1961   On being just complicated enough. Proceedings, National Academy of Sciences 47:458–464.

WEINREICH, U.
    1953   Languages in contact. New York: The Linguistic Circle of New York, Publication #1.

GESTURES: A WORKING BIBLIOGRAPHY

by
FRANCIS HAYES

GESTURES: A WORKING BIBLIOGRAPHY

by

FRANCIS HAYES

FOREWORD

IN THE ORIGINAL PLAN for this bibliography only "folk" gestures were to be included. In my article on Gesture in the *Encyclopedia Americana* (1941), I had formulated a practicable classification of gestures into three types: *folk, technical,* and *autistic.* However, within a short time after the present work was begun, the artificiality of attempting to limit research to "folk" gesticulation alone became evident. Numerous gestures have a way of jumping into two, or even three, categories. When Uriel recognized Satan by his gestures as the Fiend journeyed toward Eden, in Book IV of *Paradise Lost,* or when Moses lifted his hands on high and held them there to bring victory to the Israelites (he grew weary and had to have help), or when the minister stands, closes his eyes, lifts his face to heaven, and raises his arm to say the benediction, are these "folk" gestures? How should we classify the *gestural etymons* described by Tchang-Tcheng-Ming (*q. v.*) in his book on archaic Chinese? The lifted hand of the President of the United States at his inauguration? What kind of gesture is the salute to the flag? A Jehovah's Witness in 1943 received a life sentence (later rescinded) for *refusing* to make this gesture. Where should one put the Nazi tribal salute of hateful memory? The meanings of numerous autistic facial expressions studied by psychologists and psychiatrists are as widely understood as the handshake or the embrace; hence they fit into more than one category.

Perhaps all types of gesticulation are allied. At least one book noted in the present bibliography (Ruesch and Kees) claims there is relationship among *all types of nonverbal communication.*

Classified or not, gesture study is on the march. H. L. Mencken, the most notable American linguist interested in gestures, collected a considerable mass of material and planned an extensive section on gesticulation in a projected but never completed third supplement of "The American Language." (Personal correspondence, Jan. 3, 1944). Levette Davidson placed gesture study in its proper perspective, with reference to folklore, in *American Speech*, Feb. 1950. Maurice Krout and many other psychologists have written voluminously on gestures as they relate to the comprehension of human conduct. If we search the libraries we find acres of articles on gestures by archeologists, anthropologists, social anthropologists, sociologists, psychologists, social psychologists, linguists, folklorists, psychiatrists, teachers of the deaf and dumb, students of the American Indians, decipherers of the Maya hieroglyphics and many others. They are collecting gestures out of the past and present, interpreting them, devising a kinetic alphabet for them (Birdwhistell), searching them for the origin of numbers and counting (Dantzig, Lemoine, etc.), for the origin of writing (Danzel, Hoffman), of grammar (Goldberg), and indeed, for the origin of language itself (see especially Johannesson and Paget). As the word *significance* itself would indicate, communication by gestures almost certainly antedates communication by speech.

The present listing is a *working* bibliography, not an exhaustive one. The printing presses of today are bigger, better, faster, and almost beyond number. Students of gesticulation multiply so fast that a part-time bibliographer becomes progressively short of breath keeping pace with them. Exhaustive bibliographies compiled by any but full-time bibliographers may be a thing of the past. Withal, an exhaustive bibliography for gestures might very well be prolix beyond the point of practical use and out of proportion to the importance of the subject. Those who choose to narrow down to one topic from a field of study are still in search of gloves for handling the nettles of "the one and the many."

ACKNOWLEDGEMENTS

Special thanks are due to Dean L. E. Grinter and Acting-Dean F. W. Conner of the University of Florida Graduate School for grants which enabled me to travel to Washington and New York; to Stanley West, Director of Libraries, University of Florida; to staff members of the Library of Congress, the New York Public Library, the Columbia University Library and the University of Florida Library; to A. C. Morris, Editor, *Southern Folklore Quarterly;* to W. C. Sturtevant, of the Smithsonian Institution, for items on the sign of the fig; to Mrs. A. Hyatt Verrill, of Chiefland, Florida, for suggestions about Maya gestures found in the Maya glyphs; to Paul Williams and Curt Victorius of Guilford College; to Linton Satterthwaite, of the University of Pennsylvania Museum; to J. E. S. Thompson of the Carnegie Institution; to S. E. Leavitt of the University of North Carolina; to R. S. Boggs of the University of Miami; to Dwight Bolinger of the University of Southern California; to J. H. Groth of the University of Florida; to Lutz Röhrich, of Mainz, Germany, for numerous items printed in European periodicals; to correspondents who, unasked, kindly sent in items for this bibliography.

Particular acknowledgment is due Archer Taylor of the University of California at Berkeley. Without his help, tangible and intangible, the bibliography would never have reached publication.

Errors of fact or judgment are the compiler's alone.

COMPILER'S NOTE

The following abbreviations are placed after numerous, but not all, items to indicate at least one library where they may be found.

LC—Library of Congress, Washington, D. C.

NYP—New York Public Library, 42nd St. & Fifth Ave., New York City.

Col. Univ. Lib.—Columbia University Library, New York City.

Fla.—University of Florida Library.

Items cited without comment have not been seen by the compiler.

The compiler's definition of a gesture in the *Encyclopedia Americana* is as follows: "Any bodily movement excepting that of vocalization made consciously or unconsciously to communicate either with one's self or with another." Additionally we may include such semantic acts as "throwing down the gauntlet," or chauvinistically waving a flag, wherein something besides the body is needed to make the gesture complete.

F.C.H.

Abel, Carl. Linguistic essays, London, 1882. p. 265.
> "L'importance de la gesticulation dans l'entente des langues non-civilisées a été clairement et irréfutablement mise au jour par Carl Abel. . . ." —van Ginneken, p. 11.

Adams, Mrs. Florence Adelaide. Gesture and pantomimic action. Albany, N. Y., 1891, 221 p. Illust. LC.
> One of the most elaborate expositions of codified gesture. An actor's manual; 19th century style of dramatic expression. The "philosophy" sprinkled here and there seems quaint.

Adoration (Gestures of). Psalms CXXXIV, 2; CXLI, 2.

Aeppli, F. Die wichtigsten Ausdrücke für das Tanzen in den romanischen Sprachen. Beiheft zur Zeitschrift für romanische Philologie, vol. 75. Jahrgang, 1925. Fla.

Agreement (or pledge). "Bouvard was tired out. He let everything go for a sum so contemptible that Gouy at first opened his eyes wide, and exclaiming, 'agreed!', *slapped his palm.*" Flaubert, Bouvard et Pécuchet, chap. II (near end).
> A pledge is made by a handshake in Lope de Vega (?), La estrella de Sevilla, II, 1.1525.

Agrippa von Nettesheim, H.C. Three books of occult philosophy. Chicago, 1898. See chapter L11 "Of the countenance and gesture," et passim.

——Magische Werke. Berlin, 1916. 5 vols.
> Gestures and astrology, vol. I, p. 234 ff.; II, 93, c.16; *passim*.

Alberti, E. A. A handbook of acting. New York, 1933, 205 p.

Allen, James Turney and Italie, Gabriel. A concordance to Euripides. Berkeley, 1954, 686 p. In Greek. Fla.

Allen, John Romily. Early Christian symbolism in Great Britain and Ireland before the 13th century. London, 1887, xix, 408 p.
> Illustrated. P. 162-163, pictures of "the Almighty hand of power and protection" making the gesture with thumb, index and middle finger out, other two fingers folded. (So sacred apparently to Canon Jorio that he omitted it from his book).

Allport, F. H. Social psychology. Boston, 1924. xiv, 453 p. See p. 226 et passim.

Allport, G. W. and Vernon, Phillip. Studies in expressive movement. N. Y. 1933, 269 p. LC, FLA.
> The psychologist's approach to gesture, gait, and numerous other expressive movements. Offers a "tentative classification of expressive movement."

Amira, Karl von. Die Handgebärden in den Bilderhandschriften des Sachsenspiegels. Abhandlung der philosophisch-philologischen Klasse der königlichen bayerischen Akademie der Wissenschaft. Vol. 23. München, 1909, p. 163-263. One page of illustrations of hand gestures. NYP
> Heavily documented study of REDEGEBÄRDEN, HINWEISENDE GEBÄRDEN, DARSTELLENDE GEBÄRDEN, TAST- UND GREIFGEBÄRDEN. Bibliography in footnotes.

Ammann, Hermann. Die menschliche Rede. Sprachphilosophische Untersuchungen, 2 Bände, Lahr i.B., 1925-1928. LC
> Contents: I. T. Die Idee der Sprache und das Wesen der Wortbedeutung. - II. T. Der Satz: Lebensformen und Lebensfunktionen der Rede. Das Wesen der Satzform. Satz und Urteil.

Ammann, Johann Conrad (1669-1724). Dissertatio de loquela surdi et muti. Frankfurt, 1700. Translation into English: A dissertation on speech etc., by John Conrad Ammann, MD, London, 1873.
> A much reprinted book on lip-reading and learning speech, for the deaf and dumb.

Anderson, J. D. The language of gesture. Folk-lore, vol. XXXI (1920), p. 70. (Bengali gestures). Fla.

Anderson, Jack. 'The evil eye' and hexing. West Virginia Folklore, III, p. 42-43. (About an Austrian hexing.)

Andree, Richard. Ethnographische Parallelen und Vergleiche. Leipzig, 1878 und 1887. ('Böser Blick" p. 35-45); in 1889 ed., see chap. entitled "gemutsäusserungen und gerärden." Very few illustrations. LC

Anthriotis, N. P. Ancient and modern Greek hand and facial gestures, Morphais (Thessaloniki) (February, 1947), p. 90-92. (In Greek).

Appel, G. De Romanorum precationibus. Religionsgeschichtliche Versuche und Vorarbeiten, VII (1909), p. 2.

Approval (Russia). After a speech in Russia, Bertrand Wolfe was tossed in the air several times by several men as others stood around and applauded.

Apuleius. The golden ass, translated by Robert Graves. New York, 1953. See chap. 11, p. 150: kissing.

Arabian nights: see Book of the thousand nights and a night, The.

Aristotle: see H. Bonitz, Index Aristotelicus. Graz, 1955. (In Greek).

Arnals, Alexander d'. Der Operndarsteller; Lehrgang zur musikalischen Darstellung in der Oper. Berlin, 1932. Illust., see chapter VII on gestures used in acting in opera. LC

Árnason, Jón (1665-1743). Dactylismus ecclesiasticus; Edur, fingrarím, Vidvikjandi Kyrkju-Arsins Tímum . . . Kaupmanna-Høfn: Utgefid af P. Jónnssyni, 1838, 256 p. Tables. Illust. Sign language: Iceland. NYP

Asia magazine, vol. XXVI (1926), No. 4, p. 320. One photo. In Tibet, customary greeting to a fellow traveller: thrust up thumb of right hand and thrust out the tongue.

Asiatic society of Bengal, journal, vol III, p. 619.
> Information on Persian Gestures.

Au, Hans von der. Das Patschen im Volkstanz des rhein-mainischen Raumes. Deutsche Liederkunde, Band 1, Potsdam, 1939.

——Uber Tanznamen im Rhein-Main Gebiet. Hessische Blätter für Volkskunde. Vol. 37 (1939), p. 172-184. Not illustrated. NYP
> Beitrag zur Systematik westmitteldeutschen Volkstanzgutes nach seinem sprachlichen Wesensteil.

Aubert, Charles. The art of pantomine, N. Y., Henry Holt, 1927. Translation from the French ed. of 1901. 196 diagrams.
> An actor's handbook of gesticulation: analysis of the movements of the hands, expressions of the head, etc. Says "the language of dumbshow is universal." Diagrams some French gestures unknown in America. LC

Aungier, G. J. A History of Antiquities of Lyon, Monastery in the Parish of Isleworth. London, Nichols, 1840. See p. 405-419 for "A table of signs used during the hours of silence by the sisters and brothers in the monastery of Lyon." LC

Ausfeld, C. De Graecorum precationibus quaestiones. Jahrbücher für klassische Philologie. Supplementary vol. 28 Leipzig, p. 503-547; 903.

Austin, Gilbert. Chironomia; or a treatise on rhetorical delivery: the proper regulation of the voice, the countenance and gesture. With an investigation of the elements of gesture, and a new method for the notation thereof. London, 1806. 12 engr. plates, each containing many figures. 596 p.
> Poorly organized. Principally devoted to oratorical gestures.

Austin, Mary. Gesture in primitive drama. Theatre arts magazine, vol. 11 (August 1927), p.p594-605. Illust.
> "Author draws comparisons between American Indian gestures and those of Greek and other ancient peoples, particularly in their drama, ceremonial communication with the powers of the spirit world, and dramatic gestural abstractions. The older a play is, the greater the dependence upon rite, symbol, and gesture. The pre-speech medium of drama was gesture. The true primitive knows nothing of gesture as idiosyncrasy. He never 'fidgets'. The fidgets and flourishes of our modern necessity are the half-remembered tag ends of the sign language once universally prevailing among mankind."

Bacon, A. M. A manual of gestures. 7th ed., 1891. (For orators).

Baden, T. Bemerkungen über das komische Gebärdenspiel der Alten nach den Originalen, Neue Jahrbücher f. Phil., Suppl. I (1831), S.447-456. No illustrations. LC
> Author bases his remarks on diligent observation of theatrical sources which he feels have heretofore been neglected.

Bailey, Flora L. Navaho motor habits. American anthropologist, vol. 44 (1942), p. 210-234.

Baker, Frank. Anthropological notes on the human hand. American anthropologist o.s., vol. 1, No. 1 (1888). LC
> The placing of a dead person's hand on parts of one's body to cure disease, and other beliefs about the magic in a dead person's hand.

Balogh, J. Unbeachtetes in Augustins Konfessionen. Didaskaleion (Torino), vol. 4, 1926, p. 10-21.
> Concerning ancient Christian prayer-wailing.

Baptismal gesture, Look magazine (March 10, 1942).
> (Photograph with caption).

Barber, Jonathan. A practical treatise on gesture, chiefly abstracted from Gilbert Austin's Chironomia, Cambridge, 1831. Adapted to the use of students of Harvard. Illustrated. LC
<small>A complicated system of gestures for the orator.</small>

Barrois, J. Dactylogie et langage primitif restitués d'après les monuments. Paris, Firmin Dido, 1850.
<small>Believes in Tower of Babel. Theorizes on origin of languages from Adam down thru Noah, Moses, etc. Says Jehovah talked to the Jews with manual gestures. (See Deut. VI, 8; Exodus XIII, 9). Barrois is on steadier ground on p. 58 where he cites a list of the cultural remains of Assyrians, Indians, Egyptians, Chinese and Mexicans speculating that these people in the beginning gesticulated before they spoke. Vischnu, oldest god of India, has 4 arms and is represented as gesticulating with all of them. Planche II contains pictures of conjectured dactylogical origin of the alphabet of the Phoenicians.</small>

Bartlett, John. A complete concordance of Shakespeare. New York, 1953. See under embrace, kiss, hand, etc.

Baseball: See Signals.

Basto, C. A linguaguem dos gestos em Portugal. Revista lusitana, vol. 36 (1938), p. 5-72.

Bastow, A. Pleasant customs and superstitions in 13th century Germany. Folk-Lore, vol. 47 (1936), p. 313-28. See p. 318-9 on Fusstritt (foot-treading). Bride and groom try to tread on each other's foot: first who trod would be the master in the house. Once used as a fief symbol also.

Bateson, Gregory and Mead, Margaret. Balinese character. A photographic analysis. New York, 1942. Hand postures in daily life, p. 96. Hand postures in dance, p. 99. Prayer gesture p. 81, 229, etc. Other gestures and postures, of which there are many, passim. Bibliography, p. 255-256.

Baudin, M. Le visage humain dans la tragedie de La Calprenède. Modern language notes, vol. 45 (Feb. 1930), p. 114-119. Bibliographical footnotes.
<small>Author maintains La Calprenède pioneered in French classical tragedy in making effective use of facial gesture for dramatic effect.</small>

——Une source du décor de Racine. Modern language notes, vol. 48 (Dec. 1933), p. 501-5. Bibliog. footnotes.
<small>Author maintains Racine borrowed idea of using the human face as "un décor et un moyen d'action" from the ancients.</small>

Bauer, G. L. Die körperliche Haltung während der eucharistischen Opferfeier. Linzer theologisch-praktische Quartalschrift, vol. 89 (1936), p. 1-16.

Bauer, Leonhard. Volksleben im Lande der Bibel. Leipzig, 1903. See p. 249-256 for description of 48 Levantine gestures.

—— Einiges über Gesten der syrischen Araber. Zeitschrift des deutschen Palaestina-Vereins, Leipzig, 1898, p. 59-64. NYP

Bauer, Paul. Die Sprache der Hände; eine Einführung in die vernunftgemässe Deutung. Stuttgart, 1950. 116 p. Illust. 27 plates.

"Die Gesten der Hand," p. 14-16. A book on reading "character," etc. in the hand. "Wie sollten wir also daran zweifeln, dass unsere Hände etwas aussagen über unsere Seele, über unseren Charakter?" p. 8.

Baumeister, August. Denkmäler des klassischen Altertums zur Erläuterung des Lebens der Griechen und Römer in Religion, Kunst und Sitte. München und Leipzig, 1885-88. Band I, S. 586-592, "Gebärdensprache in der Kunst; Gebet." Fla.

Bäuml, Franz H., of the University of California at Los Angeles, is currently preparing a dictionary of gestures with illustrations.

Bayley, Harold. The lost language of symbolism. London, 1912. 2 vols.

Primarily for use of the student of written, printed, painted, or sculptured symbols. I, p. 63, portrays hands clasped; II, p. 104, a jackal-headed Egyptian ANUBIS is apparently blessing a candidate; II, 128, comments on tongue-protruding from mouth as a symbol of wisdom in Mexico and India. In Tibet a respectul salutation is made by removing hat and lolling out the tongue.

Beard, Daniel C. The American boy's books of signs, signals and symbols. Phila. and London, 1918. 250 p. 362 illustrations.

Bede, the Venerable: see Putnam F. Jones; Jacques Paul Migne.

Beil, A. Heilige Haltungen und Handlungen. Klosterneuburger Hefte 4. Klosterneuburg, Austria, n.d.

Beinhauer, Werner. El carácter español. Madrid, 1942. (Translation of Der spanische Nationalcharakter). The chapter entitled "Algo sobre el lenguaje" contains a few general remarks on Spanish gesticulation. No illustrations.

—— Spanische Umgangssprache. Berlin und Bonn, 1930. Gestures p. 113, 131, 170 (note 48), 191, 202 ff. LC

—— Über Piropos. Eine Studie über spanische Liebessprache. Volkstum und Kultur der Romanen, VII (1934), p. 111-163. Not illustrated. LC

> On women's glances see p. 133-136. Numerous quotations from literature.

Bell, Sir Charles. The hand, its mechanism and vital endowments as evincing design. Philadelphia, 1833. xii, 213 p. Illustrated.

—— Expression: its anatomy and philosophy. New York, 1873, vi, 200 p. Illustrated.

> Perhaps "the first objective and scientific study of facial expression" — Max Thorek. Primarily for students of the fine arts. Seems dated today.

Benedekfalva, M. L. de. Treatment of Hungarian peasant children. Folklore, vol. 52 (June 1941), p. 101-119.

> On the evil eye and its treatment see p. 108-114. The evil-eye is considered the greatest calamity that might threaten children. Sometimes a friend may spit in a baby's face to prove he has no "evil-eye" intentions.

Benesh, Rudolph. Introduction to Benesh dance notation. London, 1956.

> An effort to create a system of dance notation of equal utility to musical notation.

Berger, F. Körperbildung als Menschenbildung. Pädagogische Untersuchungen. V. Reihe, 44. Heft. Langensalza, 1931.

> A pedagogical psychological study.

Bergman, J. Folkloristische Beiträge. Monatsschrift für Geschichte und Wissenschaft des Judentums, vol. 79 (1935), p. 329-332.

> Concerning prayer gestures.

Berndt, R. M. Notes on the sign-language of the Jaralde tribe of the lower river Murray, South Australia. Royal Society of South Australia. Transactions and proc. and report. Adelaide, vol. 64 (1940), p. 267-272. Illust. Bibliography p. 272. NYP

Best, Harry. The deaf. New York, 1914, 340 p.

> On signs as a means of communication, see part II, chap. 19.

Bharata Muni (supposed author). Tandava Laksanam; or, The fundamentals of ancient Hindu dancing, being a translation into English of the fourth chapter of the Nātya Sāstra of Bharata. Madras, 1936. LC

> Illustrated with photos of sculptured dance and gestures posed in the great temple of Siva Nataraja at Cidambaram. Contains special appendices of aesthetic and archeological interest and a glossary of tehnical dance terms.

Bible. New Testament: Matthew 5-7. American Indian Sign Language, 1890. The Sermon on the Mount in the Indian Sign-Talk. Fort Smith, Arkansas, 1890. Illus.

——see Vorwahl; consult Harper's Bible Dictionary, by M. S. Miller and J. S. Miller, New York, 1952, under GESTURE, p. 222-223.

> These editors say that the word "gesture" is not mentioned in the Bible but indicated by scores of allusions, of which they cite thirty-six. "Easterners are given to visual expressions of emotion," they declare.
> See also MOURNING RITES; and in any Bible concordance, see under nod, beckoning silence, standing to pray, Moses holding arms aloft to win battle, kiss, knees, facing Jerusalem, etc.

Birdwhistell, R. L. Introduction to kinesics. University of Louisville, Louisville, Kentucky, 1952.

> A suggested method for cataloguing and interpreting all body movements. Author has devised a kind of kinesic shorthand, or set of shorthand kinesic pictures with which bodily movements of a subject may be quickly recorded by observer. His point of view is that of the psychologist and social anthropologist.

——"Do gestures speak louder than words?" Collier's (March 4, 1955), p. 56-57. A popular article. Illustrated.

——"Background to Kinesics." ETC., vol. XIII (1955), No. 1 p. 10-18. (ETC. is a periodical)

Blackmur, R. P. Language as gesture—essays in poetry. New York, 1935, 440 p. (Numerous later editions)

> A study of the thesis that "gesture is not only native to language, it comes before it in a still richer sense, and must be, as it were, carried into it whenever the context is imaginative." The first chapter provides a brilliant interpretation of gestures and the arts. The author's major concern is gesture and poetry, or the semantics of kinetics.

—— Language as gesture. Accent anthology, edited by Kerker Quinn and Chas. Shattack. New York, 1946, p. 467-488. Fla.

Blake, Wm. Harold. A preliminary study of the interpretation of bodily expression. New York, 1933. Ph.D. thesis. Teacher's College, Columbia University. Fla.

> A psychological, experimental approach to the interpretation of individuals by "reading" the meaning of their bodily movements.

Blessing. "The act of blessing was usually performed by the imposition of hands (e.g., Genesis XLVIII, 17-19; Matthew XIX, 13); or where a number of persons were concerned, with uplifted hands (e.g., Leviticus IX, 22; Luke XXIV, 50)." In Kings VIII, 14 and 55, the people *stand* to receive a blessing.—James Hastings, A Dictionary of the Bible, Edinburgh and New York, 1906, I, p. 307.

Blessing gesture (Hebrew, sacred); palms down, thumb nails touching, index and middle fingers touching but held separated from the ring and little fingers which touch each other. Gesture is covered by a cloth. If an ordinary person sees the gesture, he will go blind. It may only be given by high priests, descendants of Aaron. Informant: Aharon Amramy.

Blessing with extended fingers; "Y, extendiendo dos dedos, (el obispo) bendijo a la señá Frasquita y después a los demás circunstantes." P. Antonio de Alarcón, El sombrero des tres picos, Chap. XII.

Bloomfield, Leonard. Language. New York, 1933.

Blyth, John W. What is a sign? Philosophy and phenomenological research, vol. 13, No. 1 (Sept. 1952), p. 28-41.

Bodel, Jean. Le Jeu de Saint Nicholas, ed. by Alfred Jeanroy, Paris, 1925, l. 198-201:

> Li Senescaus
> Sire, bien vous croi seur les diex;
> Mais assés vous querroie miex
> Se vous l'ongle hurtïés au dent.
> Li Rois
> Senescal, n'aiés pas doutanche;
> *(le roi fait claquer son ongle sur sa dent)*
> "D'après Monmerqué-Michel (p. 167), cette forme de serment serait attribuée aux Sarrasins dans plusieurs chansons de geste (et serait encore en quelques provinces); mais ils ne donnent aucune référence." See also H. Bredtmann.

Boethius: see Lane Cooper.

Bogen, Hellmuth and Lipman, O. Gang und Charakter. Leipzig, 1931. 122 p. Psychology of movement.

Boggs, R. S. Gebärde, in Handwörterbuch des deutschen Märchens, Berlin, 1934, vol. II, p. 318-322.
> A selection and classification of some 90 gestures of scorn, fear, surprise, joy, grief, etc. from fairy tales. Mostly joy and pain are the two fundamental and characteristic emotions represented: wringing one's hands, bowing down the head, pacing the floor in anguish, burying the face in one's hands, beating the head or breast; a girl mourning kneels and tears her clothes in pieces and pulls out her hair; one person bites his thumb. Joy is expressed by clapping the hands, love by embracing and kissing, anger by shaking fist, bashfulness by downcast eyes, hungry anticipation by licking the corners of the mouth.

Bolinger, Dwight L. Thoughts on 'Yep' and 'Nope'. American speech, vol. 21 (1946), p. 90-95.
> Gestures "form as much a part of our communicative system as words and tones, and must, along with other communicative acts, be integrated into our organon of that system before we can fully know how much importance to attach to any one of the parts—in particular, whether the present all-pervasive attention to phonology is justified."

Bollhöfer, W. Gruss und Abschied in althochdeutscher and mittelhochdeutscher Zeit. Dissertation. Göttingen, 1912.

Bonifacio, Giovanni (1547-1635). L'arte de cenni con la quale formandosi fauella visible, si tratta della muta eloquenza, che non e' altro che un facondo silentio . . . Vicenza, 1616. ca. 623 p. (Errors in paging)
> Author explains meaning of several hundred gestures; gives copious notes from Dante, classic literature, and the Bible; divides gestures according to parts of body (v. g. Della barba, De gli occhi, Delle braccia, etc.); is a fervent enthusiast of gestures. No illustrations. A fundamental work.

Bonitz, H. Index Aristotelicus. Graz, 1955. 2nd edition. 878 p. Preface in Latin, text in Greek.

Book of the dead, The. Translated by Sir E. A. Wallis Budge, London, 1928. 698 p. plus index. Illustrated. Fla.
> See passim, esp. "weeping women," p. 38; a god bowing, p. 572; kneeling and supplication, p. 610; the oft-recurring raised hands (in homage), p. 440, 448-461, and elsewhere; prostration, p. 566; etc.

Book of the thousand nights and a night, The [The Arabian nights]. Transl. by Richard F. Burton, London, 1886, 10 vols. Illust. Privately printed. Indices and appendices, vol. X, p. 261-473.
> See esp. Index II, 278-347, under kiss, kneeling in prayer (exclusively Christian), hands, beckoning (the Eastern fashion the reverse of Western), Woman (head must always he kept covered); etc.

Boring. E. G. and Titchener, E. B. A model for the demonstration of facial expression. American journal of psychology, vol. 34 (1923), p. 471-485. Diagrams.

    Report on a demonstration model for facial expression constructed so that it may be readily available for class work. A wide range of facial expressions are obtained by the interchange of a number of mouths, eyes, brows, and noses (See also Piderit; Buzby; Jarden and Fernberger).

Born, Wolfgang. Fetish, amulet, and talisman. Ciba symposia, vol. 7 no. 7 (1945), Basle. Fla.

Bouillard, G. Notes diverses sur les cultes en Chine. Les attitudes des Bouddhas. Peking, 1930.

Bowditch, Chas. P., et al. Mexican and Central American antiquities. Washington, 1904, 682 p. Illust. Fla.

Bowers, Faubion. The Dance in India, New York, 1953. Fla.

Bowers, Robert H. Gesticulation and Elizabethan acting. *Southern Folklore Quarterly*, XII (1948), p. 267-277. (See this article for further bibliography of studies on gesticulation in Elizabethan times.)

Bowing. See Genesis XXXIII, 3; XLI, 42-43; Joshua XXIII, 7. See also a Biblical concordance, or under *bow* in Hastings.

Bowing (forced). Taipei, Formosa, Jan. 27.—The Chinese Nationalist government has ruled that any government employee refusing to bow before a portrait of Sun Yat-sen, founder of the Chinese republic, is liable to punishment—presumably dismissal from his government job.—Tampa Morning Tribune. Jan. 28, 1957.

Boyvin de Vavrouy. La Physionomie ou des Indices que la Nature a Mis au Corps Humain, par où l'on peut découvrir les Moeurs & les inclinations d'un chacun: Avec un traitté de la Divination par les palpitations, & un autre par les marques naturelles. Le tout traduict du Grec d'Adamantius & de Melampe. Paris, Louis de Vandosme, 1636. First edition, contemporary vellum, 8vo.

    A curious book, supposed to have been translated according to the title, by a boy of twelve; on the various signs of human expression, and other subjects.

Bredtmann, Hermann. Der sprachliche Ausdruck einiger der geläufigsten Gesten in den altfranzösischen Karlsepen. (Dissertation) Marburg, 1889.
> On the Mohammedan gesture of striking teeth when taking an oath, see p. 68-70.

Brewer, E. Cobham. Dictionary of miracles. Philadelphia, 1884(?). See Touching for the king's evil (p. 306ff.),Waters divided by sign of the cross (p. 337), Prayer (p. 440ff.), etc. Consult index.

Brewer, W. D. Patterns of Gesture among the Levantine Arabs. American anthropologist, vol. 53 (1951), p. 232-237.
> A number of gestures described and classified. Comments on gesture as a form of social behavior.

Broadbent, R. J. A History of pantomine. Simpkins, Marshall, 1901. "The best-known general history. Covers the Harlequinade and the English pantomime."—John Dolman, Jr.

Brockhaus, Der Grosse. Wiesbaden, 1952ff. 12 volumes. Illustrated. See Volume 4, Gebärde, Geste. Bibliography.

Broeg, Bob. Signals . . . the secret language of baseball. The Gillette Co., Boston, 1957. Illust.

Brouwer, P. C. de, Het handgebaar bij de Romeinen, Tijdschritvoor Taal en Lettern. Vol. VII, part 4.
> Gesture among the Romans.

Brown, L. P. Cosmic hands. Open court. Vol. 33 (January 1919), p. 8-26.

——Cosmic eyes. Open court. Vol. 32 (November, 1918), p. 685-701.

Brown, Moses True. The synthetic philosophy of expression as applied to the arts of reading, oratory, and personation. New York, 1886. (Elocution; gesture) LC

Bruyne, L. de. L'imposition des mains dans l'art chretien, Riv. di arch. crist. Vol. XX (1943), p. 41-153. (Cited by Enciclopedia Cattolica under Gesti)

Bühler, Karl. Ausdruckstheorie; das System an der Geschichte aufgezeigt. Jena, 1933, viii, 244p. LC
> Expression; physiognomy; gesture. "Der rhetorische Gebrauch von Mimik und Gesten nach Quintilian übertragen ins Deutsche von dr. Bruno Sonneck": p. 227-235.

Bulwer, John. Chirologia: or The natural language of the hand. London, 1644, 2 vols. Illustrated.
> Rich in historical references to gesture, especially among the Greeks, Romans, and Hebrews. Excellent for dating many gestures; references prove them already old in ancient times.

Burke, Thomas. The Streets of London. London, 4th edition, 1949, p. 94. "Crude gestures have always been a part of the London scene. In Tudor times the invitation to quarrel or combat was given by a biting of the thumb; in the middle eighteenth century by cocking the hat; later by a jerk of the thumb over the left shoulder, implying illegitimate birth; in the early nineteenth century, by the thumb to the nose, and within living memory by two fingers jerked upwards."

Burton, M. Eyes have it. Illustrated London News, vol. 223 (November 28, 1953), p. 886.

Burton, Sir Richard Francis. The city of the saints. 1862, 574p. List of Indian signs, with explanations, p. 150-160; signs not illustrated. LC

Butler, Samuel. Hudibras. London, 1907.
> Part I, Canto 2, line 681: clapped on sword to show he meant his word.
> I,2,1.737-739: "At this the Knight grew high in wroth, and lifting hands and eyes up both, Three times he smote on stomach stout."
> I,2,1.997: wink of eye, sentence of death.
> I,3,1.815: "I scorn, quoth she, thou coxcomb silly, clapping her hands upon her breech."
> II,1,1.118-20: To b'seen by her in such a place . . . made him hang his head and scowl, and wink and goggle like an owl. . ."
> II,1,1.169: Beard-pulling implied.
> II,1,1.235, note: Slave freed by a blow.
> II,1,1.540, note: One bargainer strikes other's palm, seals bargain [Still current in France].

Buzby, D. E. Interpretation of facial expression. American journal of psychology, vol. 35 (Oct. 1924), p. 602-604.
> Statistical approach to the study of gestures as represented on a demonstration model for facial expression. The model is an artifact and varying expressions are obtained by the interchange of a number of mouths, eyes, noses and brows.

Caballero, Ramón. Diccionario de modismos. Madrid, n.d.
> Contains nearly 300 Spanish phrases in which the hand figures. These phrases are put together into a short prose piece [technique resembles Blasco de Garay's in Carta de refranes, 1540]; following this is a list of all the phrases used in the piece. See p. 1181-1198.

Callisen, S. A. The evil eye in Italian art. The art bulletin (University of Chicago), XIX (1937), p. 450-462.
> On protections against the evil eye in Italian art. See summary by W. L. Hildburgh, Folk-Lore (London), vol. 49 (1938), p. 295-297.

Callotto Resuscitato, Il, Oder Neu Eingerichtetes Zwerchencabinet, cum priv. S.C. May, Augsburg, ca. 1720 (Originally Le Monde Plein de Fols, ou Le Théâtre des Nains, en François et en Hollandois, Amsterdam, 1715) See gesticulation portrayed in engravings of grotesque figures. LC

Calmet, Augustin. Dictionary of the Holy Bible. New York, 1837, 8th ed. Vol. IV, plate 35.
> Pictures a hand as FINIAL of Roman standard. LC

Campbell, A. J. D. Introduction to de Tyra Af Kleen's Mudrās. See Kleen, Tyra Af. LC

Cardona, Miguel. Gestos o ademanes habituales en Venezuela Archivos venezolanos de folklore. (Caracas, Universidad Central de Venezuela), 1953-1954, año II-III, tomo II, número 3, p. 159-166. 22 illustrations. Fla.
> Various gestures from Venezuela.

Carlile, John S. Production and direction of radio programs. New York, 1949, p. 291-300. LC
> Reproduction, well illustrated, of 25 studio gestures from Variety radio directory, vol. I, p. 328-337.

Carmichall, L. and others. Study of the judgment of manual expression as presented in still and motion pictures. Journal of social psychology, vol. 8 (Feb. 1937), p. 115-142. Illustrated. Bibliography. Tables.

Carr, Archie. The windward road. New York, 1956.
> "Chepe jerked his head toward the hut and pointed with his lower lip. 'That's where the mosquitoes are.'" P. 175.
> The author tells me that this is the usual way to point out objects in Costa Rica and other Central American countries. Pointing with the finger is tabu. In Kenya, British East Africa, the lips are similarly used for pointing.

Carus, C. G. Symbolik der menschlichen Gestalt. Radebeul-Dresden, 1939. First edition, Leipzig, 1852.

Castro, Guillén de. Las mocedades del Cid (1st play), II, lines 1239-1241: on kissing and the honor code.

Catholic encyclopedia. New York, 1912 and later. See rites, ritual, mass, genuflection, prayer (Posture), etc.

Celestina, The. Translated from the Spanish by L. B. Simpson, Berkeley, 1955, Act VI, p. 69: "she gave herself a great slap on the forehead, like one who hears a dreadful piece of news...."

    In Act V, p. 61, Sempronio, a servant, makes the sign of the cross on meeting Celestina in the street.

Ceremony: see Ritual.

Cervantes, Miguel de. Don Quixote: embracing and kissing, 2nd part, Chap. XXXV; gesture of indigence, 2nd part, Chap. LIV; bowing, 2nd part, Chap. XXXV.

Chapman, Ashton, Watch your gestures. She, Vol. I (Oct. 1943), p. 24-26.

    A journalistic article giving advice to women about folk gestures, nervous mannerisms, etc.

———Card-cataloguing man's gesture. Profitable hobbies, Vol. V (Nov. 1948), p. 30 ff.

    A popularized article.

Chase, Stuart. The language of nods. Saturday review, XL (Mar. 2, 1957), p. 17-18. An article-review of Ruesch and Kees, Nonverbal communication. Chase's article is brief and popular in style.

Churchill, J. C. Do your gestures give you away? Woman's home companion, July, 1951, p. 14. Illustrated. A journalistic article.

Cicero. De Oratore, book III, lix.

Cid, The. The Cid, Ruy Diaz de Vivar, grasped his beard to express lively satisfaction. R. Menéndez Pidal ed., Cantar de mio Cid, Madrid, 1945, Vol. II, 3rd part, p. 494.

Clapping Hands: Ezekiel, XXV, 6-7; Lamentations II, 15; Job XXXIV, 37 and XXVII, 23; Psalms XLVII, 1.

Clark, Wm. Philo (Captain). Indian sign language, Washington, Phila,. 1885, 443 p.

    Based on six years residence among the American Indians. Also provides brief explanatory notes of the gestures taught to deaf mutes. During the

Sioux and Cheyenne war of 1876-7 the author was in command of some 300 friendly Indian scouts (Pawnee, Shoshone, Arapahoe, Cheyenne, Crow, Sioux). The six tribes had six different vocal languages but a common gesture language. The author learned the gesture language and used it constantly among many tribes. Gestures listed under key word and described. No illustrations.

Clasping the knees. "Odysseus cast his hands about the knee of Arete. . . ."—Odyssey, VII, between lines 124-153.

" 'To clasp the knees was a sign of submission adopted in earnest supplication; found in Homer and in Herodotus, IX, 76 . . . Achilles clasps his mother Thetis by the knees when begging her to intercede for him (Iliad)." —From letter from J. P. Harland, Chapel Hill, N. C.

Clemen, C. "Gebärde," Die Religion in Geschichte und Gegenwart, 2nd edition, Tübingen, 1927-31, p. 868. Largely liturgical; one column of text.

Clodd, E. The story of the alphabet. New York, 1900 et seq. Illustrated with numerous drawings, many of which show ancient and primitive picture-writing and ideographs based on gestures. LC

Cobb, Jane. Clappers and hissers. New York Times Magazine, (April 21, 1940), p. 7.

Cocchiara, Giuseppe. Il Linguaggio del Gesto. Torino (Bocca), 1932, 131 p. (See Zeitschrift für Volkskunde, vol. 44, p. 279, for short and instructive review.) LC

Well annotated. Bibliographical notes at end of each chapter. Very few illustrations. Chapters are as follows: I Introduzione; II Psicologia del Gesto; III Il Gesto al principio. IV. I gesti di aggregazione; V. I gesti di neutralizzazione; VI. La funzione della mano nel gesto; VII. Le due lingue; Appendice: Di una nuova interpretazione data al linguaggio del gesto.
"Dico subito che col mio libro io non ho inteso compilare un vocabulario dei gesti nè svolgere su di essi una teoria completa e sistematica. [Mio libro] vuole essere un'introduzione alla grammatica dei gesti . . ."—From the Preface.

Cody, Iron Eyes. How: Sign talk in pictures, by Iron Eyes Cody assisted by Ye-was. Illustrated by Clarence Illsworth; posed by Iron Eyes and Ye-was. 1st edition Hollywood, California, as a Boelter Classic by H. H. Boelter Lithography, 1925. Unpaged. Illustrated. LC

Cohn, Paula. Gestures found in Grimm's Household Tales. Unpublished study made in class of Archer Taylor, University of California, Berkeley. The variety of gestures found is small. Of 300 tales, 33 contain one or more gestures.

Coleman, Charles. The Mythology of the Hindus with notices of various mountain and island tribes inhabiting the two peninsulas of India and the neighbouring islands; and as Appendix comprising the minor avatars, and the mythological and religious terms, etc. etc., of the Hindus. London, 1832. xvi, 401 p. LC

> Profusely illustrated with plates with detailed explanations of actual or conjectured meanings of numerous gestures, facial, manual, etc. Many gods and goddesses have multiple arms and hands gesticulating. Peoples included, in addition to the Indians of India, are the Japanese, the Egyptians, the Daya of Borneo, and many more.

Combs, Homer C. and Sullens, Z. R. A concordance to the English poems of John Donne. Chicago, 1940. Fla.

Comstock, Andrew. A system of elocution, with special reference to gesture, to the treatment of stammering, and defective articulation. Philadelphia, 1858. 20th edition. Fla., LC

Concordances. Concordances of literature of the past afford description of numerous gestures. See, for example, Ebeling, Concordance to Homer; James Allen (Euripides); Homer Combs (John Donne); Lane Cooper (Boethius); Putnam Jones (Bede); John Bartlett (Shakespeare); and the latest Concordance of the Bible. There are others. They afford one of the best means for ascertaining the approximate age of specific gestures.

Conway, M.D. The oath and its ethics. London, 1881. 27p. NYP, LC

———. Demonology and devil love. New York, 1879, 2 vols. Illustrated. LC

Cook, Arthur Bernard. Cykøanthc. Classical Review (London), vol. 21 (1907), p. 133-136. Illust. Meaning of the sign of the fig used to clear up obscure Greek word. Fig sign also found in Aristophanes Comoedias, ed. by Theodorus Bergk, Lipsiae, vol. 1, 1903, Pax, line 1350.

Coomaraswamy, A. and Gopalakrishnāyya (translators). Nandikesvara, The mirrour of gesture. 2nd edition, 1936. 81 p. 20 pl.

> Excellent description of technical gestures found in dancing of India. Good illustrations and bibliography.

Cooper, Lane. A concordance of Boethius; the five theological tractates and the Consolation of philosophy. Cambridge, Massachusetts, 1928.

Cooperation in a silent world. Manchester Guardian Weekly, vol. 77 (August 15, 1957), p. 14.

> Author (anonymous) visits for a week with a deaf family in Essex. Among other comments on sign language, he says a deaf Frenchman, only three days in England, was able to communicate with English deaf mutes; their language in pictures made communication easy. However, "like our spoken language there are dialects and national sign languages."

Cosgrove, D. A study of the reliability of judging emotions as expressed by the hands. University of Detroit, 1954. Unpublished master's thesis.

Cossetta, Al. Natural gestures and postures in speech. Kansas City, Missouri, 1946. 151 p. Illustrated. LC

> This book is published by the Natural Gestures and Postures Association, 1200 Oak Street, Kansas City 6, Missouri. Exemplary quotations: "Many books devoted to the subject of public speaking have been written, but none are compiled with illustrations of the right gestures . . . (sic) . . . All of us are salesmen. . . . Believe in what you have to 'sell' and you will 'sell' it." (sic) Not recommended.

Cough (as gesture).

> "The preachers . . . looked upon coughing and hemming as ornaments of speech; and when they printed their sermons, noted in the margin where the preacher coughed or hemm'd. This practice was not confined to England, for Oliver Maillard, a Cordelier, and famous preacher, printed a sermon at Brussels in the year 1500, and marked in the margin where the preacher hemm'd once or twice, or coughed." Hudibras, Samuel Butler, London, 1907, Part I, Canto 1, lines 81-86. Note of editor, Henry G. Bohn.

Counting: see Dantzig; Lemoine; Tchang Tcheng Ming.

Covering head. "Any man who offers prayer or explains the will of God with anything on his head disgraces his head, and any woman who offers prayer or explains the will of God bareheaded disgraces her head, for it is just as though she had her head shaved."—I Corinthians XI, 4-6.

Craig, Alice E. The speech arts; a textbook of oral English. Revised ed., New York, 1941. xix, 610 p. Illustrated. LC

> See chapter on "Pantomime and gesture." Includes tables, diagrams, forms, bibliographies.

Critchley, MacDonald. The language of gesture. London and New York, 1939. 128 p. Not illustrated.

> A scholarly study in 18 chapters. Some of the chapters are entitled as follows: "Gesture and speech," "The sign-talk of deaf mutes," "Gesture as a precursor to language."

——— Brain, vol. 61 (1938), p. 163. On the neurology of the gestures of a partial deaf-mute.

——— and Earl, C. J. C. Brain, vol. 55 (1932), p. 311. On postures inexplicably assumed by schizophrenics.

Cuisinier, Jeanne. The gestures in the Cambodian ballet: their traditional and symbolic significance, Indian art and letters, New series, vol. 1, (London, 1927) p. 92-103. Bibliography, p. 103.   NYP

Cushing, J. H. Manual Concepts. American anthropologist, vol. 5 (1892), p. 289-317. Illustrated.

> The gist of the article is in its subtitle: "A study of the influence of hand-usage on culture-growth." Author believes that the hand of man is so intimately associated with the mind of man that the hand "moulded intangible thoughts no less than the tangible products of his brain." He cites evidence of manual influence in the formation of spoken language and in the invention of "a cumbersome decimal system of enumeration when . . . a duodecimal system would be better." He supports his thesis by citing examples from the Zuñi language, and Roman and Chinese numerals.

Cuyer, Édouard. La Mimique. Paris, 1902, 366 p. (Bibliothèque Internationale de Psychologie Expérimentale) Bibliography p. 355-7. Brief dictionary of gestures p. 307-351. Illust.

> Numerous folk gestures discussed; author's point of view is that of experimental psychologist. He wishes to learn "comment les émotions se traduisent à l'extérieur, et surtout pour-quoi elles se traduisent d'une certaine façon."

Daniel-Rops. Les gestes de la prière et l'imploration, France illustration, vol. 7, Dec. 1, 1951, p. 599-606.

Dante. The divine comedy. Translated by H. R. Huse. New York, Rinehart and Co., 1954.

> See Purgatory, Canto 21: a smile and an embrace; Inferno, Canto 25: Vanni Fucci makes sign of fig, insulting Creator.

Dantzig, Tobias. Number: the language of science; a critical survey written for the non-mathematician. 4th ed. New York, 1954. 340 p. Illustrated. Fla.

> Chapter I on the hand and counting. Illustration of finger symbols of 16th century, p. 2. A popularized survey on the whole.

Danzel, Theodor-Wilhelm. Die Anfänge der Schrift. Leipzig, 1912. Illustrated. LC

> Concerned passim with gesticulation and the origin of writing; includes writing of Mexicans, Chinese, Babylonians, Egyptians, etc.; world-wide point-of-view.

────── Codex hammaburgensis, eine neuentdeckte altmexikanische Bilderhandschrift des hamburgischen Museums fur Völkerkunde. Hamburg, 1926. Mexican Indian writing. Fla.

────── Kultur und Religion des primitiven Menschen. Einführung in Hauptprobleme der allgemeinen Völkerkunde und Völkerpsychologie. Stuttgart, 1924. viii, 132 p. Illustrated. LC

> Gestures passim, as used in magic and in picture writing, especially in Hispanic America. Bibliography of author's works to date ((1924), p. 134.

────── Magie und Geheimwissenschaft. Stuttgart, 1924. xv. 213 p. LC

> A few illustrations. Chapters on magic in ancient Mexico, Peru, ancient Egypt, China, Assyria, etc. No index for locating comments on magic gestures.

Daremberg, Charles, and Saglio, Edmond. Dictionnaire des Antiquités Grecques et Romaines. Paris, 1877 et seq., 5 vols. Illustrated. Fla.

> Rich in gestures. See, for example, prayer gestures, vol. 2, 2ᵉ partie, "Funus," p. 1367 ff.; "Forge," fig. 2964; etc. A fundamental work.

Darwin, Charles R. Expression of the emotions in man and animals. New York, 1955. xi, 372 p. Introduction by M. Mead. Illustrated.

> 19th century speculation on gestures passim.

Davidson, Levette J. Some current folk gestures and sign language. American speech, vol. 25 (Feb. 1950), p. 3-9.

> A scholarly, highly-readable article. Summarizes briefly and skillfully what gesture research is all about and places it in perspective with respect to oral and written communication.

────── A guide to American folklore. Denver, U. of Denver Press, 1951. ix, 132 p. Fla.

> See chapter 7. Rather harshly reviewed by Daniel Hoffman in the Journal of American folklore, vol. 65 p. 199.

Davis, R. C. Specificity of facial expressions; correction of a statistical misinterpretation in Landis' experiment. Journal of general psychology, vol. 10 (Jan. 1934), p. 42-58.

> Proves Landis' experiment (q.v.) on facial expression contains an error in statistical procedure which invalidates his conclusions.

Deaf and dumb: See Max A. Goldstein, Problem of the deaf. St. Louis, 1933. 580 p. Illust. Bibliog. p. 573; or K. W. Hodgson, The deaf and their problems, with a preface by Sir Wm. Paget, London 1953, xx, 364 p.; Handbuch Des Taubstummewesens, Osterwieck Am Harz: E. Staude, 1929, xi, 744 p. The New York Public Library has bound together, without cataloguing, a collection of pamphlets on the deaf.

> Gallaudet College, Washington, D. C., is a source of information on the deaf and dumb; also Deutsches Museum für Taubstummenkunde.

Deaf and dumb persons' alphabet in Pequeño Larousse. Paris, 1926, p. 861. Illust.

De Battini, Berta Elena Vidal. El habla rural de San Luis. Parte I: Fonética, morfología, sintaxis hispanoamericana. Buenos Aires, vol. 7 (1949), p. 209-213.

Deferrari, Roy J. et al, A Concondance of Ovid, Washington. 1939, 2220 p. Fla.

——— A concondance of Prudentius. Cambridge, Mass., 1932. Fla.

De Haerne, D. The natural language of signs. American annals of the deaf & dumb, 1875.

> Interesting and valuable for background in studying gestures. Discusses possible relationships between sign-lang. & origin of all lang., sign lang. & psychology, etc. Slanted toward the gestures of the deaf & dumb, & North American Indians & other heterogeneous groups who are able to speak with each other by signs.

Dekker, Thomas. Old fortunatus. Mermaid Series, London, 1949.

> The connuto gesture is prominent in Act I about last 20 lines of the play.

Delacroix, Henri J. Le language et la Pensée. Paris, 1933.

———Les Operations intellectuelles. Paris, 1936. In Nouveau Traité de Psychologie, of G. Dumas, vol. V, fascicule 2, p. 114 sq. author suggests broadening the concept of the word language.

Delaumosne, L'abbé. The Delsarte system. Translated by Frances A. Shaw. A complete explanation of the Delsarte system of gestures by one of Delsarte's pupils. Alarmingly exhaustive aesthetic talk about posture, gesture, the walk, the hand, the principle of the trinity, etc. Often obscure.

Delsarte, François (1811-1871) (See Delaumosne, L'abbé; Stebbins, Genevieve; Shawn, Ted).

Deutsches Museum für Taubstummenkunde. Leipzig, Katalog I, 1952.

Dhorme, E. L'Emploi métaphorique des noms de Parties du Corps en Hebreu et en Akkadien, Revue biblique (1921), p. 374-399; 517-540; (1922), p. 215-233, 489-517; (1923), p. 185-212.

Diccionario enciclopédico Hispano-Americano. Barcelona, 1892. On Gesto, Gesticular, etc. with a few literary references, see vol. 9, p. 366.

Dickey, Elizabeth and Knower, Franklin H. American journal of sociology, vol. 47, no. 2 (Sept. 1941), p. 190.
> A note on some ethnological differences in recognition of simulated expressions of the emotions.

Dictionaries
See under kiss, hand, arm, bow, wink, finger, brow, etc. in American, English, French, German and other dictionaries. The Diccionario de modismos of Ramon Caballero contains many phrases in which the hand figures. The following are also recommended:
Schwäbisches Wörterbuch
Schweizerisches Idiotikon
Bayrisches Wörterbuch
Rheinisches Wörterbuch
Badisches Wörterbuch
Preussisches Wörterbuch
Mecklenburgisches Wörterbuch
Deutsches Wörterbuch der Brüder Grimm
Roschers mythologisches Lexikon
Pauly - Wissowas Realenzyklopädie
See also Godefroy, Dictionnaire de l'ancienne langue française, New Oxford English Dictionary, the dialect dictionaries, etc.

Dictionary without words. Better English, VIII (June, 1942), p. 5.
> Brief journalistic piece.

Dieterich, Albrecht. Abraxas. Studien zur Religionsgeschichte des spätern altertums. Liepzig, 1891.
> Abraxas is a magic name found engraved on ancient stones. In Greek notation it equals the number 365, corresponding to the 365 spirits emanating from the Supreme Being.

Digiti lingua. London, 1698. (Item from Critchley, q.v.).

Dilthey, Karl. Archaeologisch epigraphische Mittheilungen aus Oesterreich-Ungarn. II (1877). (Item cited by Elworthy, q.v.).

Diringer, Daniel. The alphabet; a key to the history of mankind. New York, 1948. 607 p. Illust. Bibliog.
>See passim for references to gestures.

Doat, Jan. L'expression corporelle du comédien. Grenoble, 1944. 72 p. Illust. Bibliog., p. 71-72.
>An actor's manual. Author's thesis: "Nous subissons encore la néfaste influence, d'une époque qui a réduit aux minimum la notion de l'expression corporelle de l'acteur . . . Le corps humain en mouvement doit obéir à une loi d'esthétique: l'enchaînement des gestes et des positions. C'est une sorte de syntaxe . . ."

Dölger, F. J. Sol salutis. Gebet und Gesang im christlichen Altertum, mit besonderer Rücksicht auf die Ostung in Gebet and Liturgie. Münster in Westfalen, 1920, vol. 2, 342 p.
>Important bibliographical items in footnotes; references mostly classical Greek and Latin and medieval Latin. Not illustrated. Comparative liturgies, with special reference to Sonnenkult und Christentum. For gestures, see Namen- und Sachregister under Aufblick zum Himmel, Handausstrecken als Befehlgestus, Kreuzzeichen, etc. p. 327-342.

——— Der Exorzismus im altchristlichen Taufritual. Paderborn, 1909. xi, 175 p.

Döller, J. Das Gebet im Alten Testament in religionsgeschichtlicher Beleuchtung. Theologische Studien der Oesterreichischen Leo-Gesellschaft, vol. 21. Vienna, 1914.

Dolman, John, Jr. The art of acting. New York, 1949. 305 p.
>See Chapter XVI, "Bodily action;" also bibliog., p. 289.

Domenech, Emmanuel H. D. l'Abbé. Manuscrit pictographique américain, précédé d'une notice sur l'idéographie des peaux-rogues. Paris, 1860. 228 p. Illust. LC
>Interpretation and discussion of numerous reproductions of American Indian pictographs; gestures predominate in most of them. The salvaging of the manuscripts is due to the zeal of an 18th century Italian, Boturini. (The publication of this book called forth numerous articles which attacked its authenticity.)

Doncoeur, P. L'humble prière de nos corps. Des saintes extravagances des stylites à l'harmonie humaine de la prière catholique. Études, Paris, vol. 178 (1924), p. 200-217.

Dorcus, R. M. Hypnosis and its therapeutic applications. New York, 1956. Chap. II, p. 7-8. Fla.
>Describes use of hands and glance of eyes for inducing hypnotism. [Query: What is the relationship between hypnotism and the origin of belief in the evil eye?]

Douailler, Charles. Manuel technique et pratique du tragédien et du comédien tyrique. Paris, 1911. 67 p. Illust. LC

> An actor's manual. Author defends the thesis that there are laws concerning gesture and posture which an actor should learn, just as there are laws of music which a musician must know. He interprets actor's gesture à la Delsarte.

Doutté, Edmond. Magie dans l'Afrique du Nord. Algiers, 1909. 617. p. Illust. NYP

> "Les rites sont ou des gestes, ou des paroles, ou des figures."—p. 599. See especially chap. II.

Drechsler, Paul. Sitte, Brauch, und Volksglaube in Schlesien. Leipzig, 1906. 2 vols. Some illustrations. LC

> A number of Silesian gestures.

Drezinski, Herman. Gesture 'art' south of Border. New Orleans Item, April 23, 1950.

> Slight journalistic article illustrated with two photos of some Latin American gestures.

Dubois, Louis F. Histoire civile, religieuse et littéraire de l'Abbaye de la Trappe. Paris, 1824. 386 p. NYP

> See especially chap. XI, which describes a number of prayer gestures and contains an alphabetical list of Trappist gestures, each one explained (p. 248-258).

Duchesne, Mgr. L. Christian worship: its orgin and evolution. London, 1903. x, 558 p. Translated from the 3rd French ed. by M. L. McClure. NYP

> A study of the Latin liturgy up to the time of Charlemagne. See Index under kiss, cross, anointing, liturgy, etc.

Duggan, Anne S. et al. The folk dance library. New York, 1948. 5 vols. Illustrated. Fla.

Dumas, Georges and André Ombredane. Nouveau traité de psychologie. Paris, 1930 et seq. 5 vols. LC

> See especially vols. 2 and 3 for exceedingly detailed studies of gestures of pleasure, pain, mimicry; gestures & communication of ideas, etc. See also vol. 4, p. 334-336: La symbolization dans la mimique. The perspective is that of the psychologist but the conclusions drawn are valuable to the folklorist. The authors occasionally lean heavily on speculation. Long bibliographies at the end of each chapter mostly list studies of psychologists.

——— La vie affective. Paris, 1948. Illust.

> Bibliography of numerous works of Dumas given on frontispiece. LC "C'est une sorte de refonte de tout ce que j'ai publié de 1892 à 1935

sur les phénomènes affectifs . . . Dans chacun des chapitres, on a taché d'étudier les phénomènes affectifs dans leur origine . . ." The author studies gesticulation as a doctor would, but his ideas and examples cited and illustrated are stimulating and valuable to the folklorist. His work is admittedly inspired "dans ses grandes lignes du mécanisme cartésien." Some of the same material from the author's Nouveau traité de psychologie appears in La vie affective.

Dunbar, William. Wm. Dunbar's List. Publications of the transactions of the American philosophical society (Jan. 16, 1801). Fla.

About 60 signs.

Dunlap, K. A project for investigating facial signs of personality. American journal of psychology, vol. 39 (1927), p. 156-161.

Ear-wiggling philatelists. An Associated Press story for September 8, 1956. (Appeared in numerous U. S. newspapers.)

At a stamp auction, bids are made by wiggling ears, tapping chest with pencil, moving an ankle, and similar gestures.

Ebeling, Heinrich. Lexicon Homericum. . . . Lipsiae, 1885. 2 vols.

Ebermann. Zeitschrift des Vereins für Volkskunde. XXVIII (1918), p. 142.

Contains good picture of two fingers held up to ensure the validity of an oath (called die Eidfinger). See early parallel in Hartmann von Aue, cited in A. Taylor, The Judas curse, Am. journ. philol., XLII (1921), p. 251.

Echtermeyer, Th. Über Namen und symbolische Bedeutung der Finger bei den Griechen und Römern. Programm des Pädagogiums in Halle, 1835.

Efron, David. Race and gesture. Council for Research in the Social Sciences. Columbia University. (Item mentioned to me in correspondence with Efron. I have not seen it.)

——— Gesture & environment; a tentative study of some of the spatio-temporal and 'linguistic' aspects of the gestural behavior of Eastern Jews and Southern Italians in New York City, living under similar as well as different environmental conditions. New York, King's Crown Press, 1941. 194 p. Diagrams & illustrations. Bibliog. LC

A useful, although poorly organized, study of several phases of gesticulation. The author is over-immersed in esoteric statistical tables and graphs. Conclusions, p. 136-137.

———— and J. P. Foley, Jr. Gestural behavior and social setting. Zeitschrift für Socialforsch, vol. 6 (1937), p. 152-161. (New name: Studies in philosophy & social science.) LC

> Authors' researches reveal that the typical gestural patterns of Jews and Italians tend to alter or disappear with their assimilation into the American community.
> (In personal correspondence Mr. Efron told me he was going to publish a bibliography of some 1000 items on gesture & posture in May, 1942. I have been unable to locate it.)

Eisenberg, P. Motivation of expressive movement. Journal of general psychology. Vol. 23 (July, 1940), p. 89-101.

> Expressive movement alone is insufficient for revealing 'inner disposition' or the 'organization of personality.'

Eisenhofer, L. Handbuch der katholischen Liturgik I. Freiburg im Breisgau, 1932.

> On liturgical gestures, p. 251-282.

Eller E. Das Gebet. Religionspsychologische Studien. Paderborn, 1937.

Elworthy, Frederick Thomas. The evil eye. An account of this ancient and widespread superstition. London, 1895. 471 p. Profusely illustrated. Appendix. Good Index. LC

> See especially chap. VII on touch, gestures; and chap. IX, the Mano Pantea. Some duplication of the evil eye in the author's Horns of honour, q.v. An indispensable work.

———— Horns of honour. London, 1900. Profusely illustrated. LC

> See especially articles on 'Horns of honor,' 'Horns of the devil,' 'The hand,' and 'The thumb.' Rich in bibliog. But items often incomplete. Many sources cited, especially ancient and medieval. Worldwide in scope Author occasionally overly-speculative, but the work is indispensable. Duplicates parts of author's book entitled The evil eye.

———— Evil eye. Hasting's encyclopedia of religion & ethics. Vol. 5, p. 608-615.

Embrace. See Genesis, XXXIII, 4; Acts, XX, 37-38; Shakespeare, 3 Henry VI, II, 3, lines 44-5.

Emergency signals. Flying cadet (Sept. 1943). A government war publication. Illustrated. LC

> Packed with every parachute used in the South Pacific by army airmen.

Encyclopedia Cattolica. Rome, 1951.

> See under Gesti. Very brief article, and bibliography, with special attention to religious gestures.

Encyclopedia Espasa-Calpe. Barcelona, 1924 et seq. Vol. 25, p. 1508-1512.

Encyclopedia Americana. New York, 1941 et seq.
> See articles under "prayer rug" and "gesture."

Encyclopedia Italiana. Milan, 1932. Vol. 16, p. 861 & 863: gestures of Christ. (See passim for other paintings with gesture motifs.)

Encyclopédie Mensuelle d'Outre-Mer. Paris, 1954.
> During Islamic conquest, a raised forefinger of the right hand signified acceptance of Islam (p. 10).

Engel, Johann Jakob. Ideen zu einer Mimik. Berlin, 1785-6. 2 vols. Later eds., Berlin, 1804 & 1810. In French: Idées sur le geste et l'action théâtrale. Paris, 2 vols., 1788-9. LC
> An actor's manual of gestures. Illustrated.

Enjoy, Paul d'. Le rôle de la main dans les gestes de responsabilité. Revue scientifique (Formerly Revue Rose) ser. 4, vol. 14 (1900), p. 81-3. NYP

Epictetus in the Harvard Classics, New York, 1910, vol. II, p. 131.
> When a man has been raised to high political office, everyone that he meets congratulates him. One kisses him on the eyes, another on the neck, while slaves kiss his hand.

Euripides, A concordance to. See James Allen & G. Italie.

Evans, E. C. Quo modo corpora voltusque hominum auctores latini descripserint; summary. Harvard studies in classical philology, vol. 41 (1930), p. 192-195.

Faber, Albrecht. Laut- und Gebärdensprache bei Insekten. Stuttgart, 1953. 198 p. Illust.
> The sub-title reads: "Vergleichende Darstellung von Ausdrucksformen als Zeitgestalten und ihren Funktionen."

Fascist salute in Spain. See frontcover of The Spanish Information Bureau, (Jan. 1939). 110 E. 42 St., New York.

Fascist salute in Lower California. See Pic Magazine (Jan. 6, 1942).

Feet (uncovering).
> To keep on one's shoes on entering a friend's home or the temple was bad manners.—A. C. Bouquet, Everyday life in New Testament times. New York, 1955, p. 144.

Fenichell, Otto. Die 'lange Nase.' Imago. Zeitschrift für Anwendung der Psychoanalyse auf die Natur- und Geisteswissenschaften. Vol. XIV (1928), p. 502-504.

Ferdon, Constance Etz. Mind your manners. Inter-American, V (July, 1946), p. 19-20; p. 40.
> A brief journalistic piece with gestures illustrated.

Fernberger, S. W. False suggestion and the Piderit model. American journal of psychology, vol. 40 (Oct. 1928), p. 562-568. Bibliog. footnotes.
> The fourth report of a series of experiments performed with the Piderit model constructed for demonstration of facial expression.

———. Can emotion be accurately judged by its facial expression alone? Journal of criminal law & criminology, vol. 20 (Feb. 1930), p. 554-564. Bibliographical footnotes. Tables.

Ferrero, G. Les lois psychologiques du symbolisme. Paris, 1895.
> "Aujourd'hui encore lorsque nous voulons affirmer avec énergie notre droit de propriété sur une chose, même sur une chose lointaine ou immatérielle, nous tendons les bras comme pour la saisir." This gesture goes back to the time when we *took* what we desired, if we were able.
> The work, in general, is speculative.

Field, C. (Colonel). Salutes & saluting, naval & military. Journal of the royal united service institution, LXIII (1918), p. 42-49.
> Maintains that the modern military salute began in the 18th century.

Fields, S. J. Discrimination of facial expression and its relation to personal adjustment. Journal of social psychology, vol. 38 (Aug. 1953), p. 63-71.

Finger, Magic hunting. Seneca fiction. 32nd annual report, Bureau of American ethnology. Washington, D. C., vol. 32 (1910-11), p. 266.

Finger signs (used on grain exchange). National education association journal, vol. 31 (1932), p. 282. Illust.

Fingers crossed (for luck). Governor Leroy Collins (Florida), How we solve our teen-age problem. Saturday Evening Post (April 21, 1956).
> Large photo of Pensacola high school girls' choir in act of crossing fingers for luck before competing.

Fischer, Captain Harold E. Jr. My case as a prisoner was different. Life magazine, vol. 38, no. 26 (June 27, 1955), p. 157.

> Photo of Fischer and others in a Chinese prison. He and another American make secret gestures, unknown to the Chinese, but clear to American readers.

Flachskampf, Ludwig. Spanische Gebärdensprache. Erlangen, 1938. Reprinted from Romanische Forschungen, vol. LII (1938), p. 205-258.

> A scholarly study of various Spanish gestures. Reviewed in Hispanic review, vol. VIII, p. 86-87.

Flick, G. Morphologie des Rhythmus. Dissertation. Berlin, 1936.

Flögel, Karl Friedrich. Geschichte des Grotesk-komischen, ein Beitrag zur Geschichte der Menschheit. Liegniss und Leipzig, 1788. 322 p. LC

> History of grotesque & comic European entertainments, farces, Narrenfest, commedia della arte characters, die Gesellschaft der Hörnerträger zu Evreux und Rouen, etc. No index.

Florenz, K. Ancient Japanese rituals. Asiatic society of Japan (Transactions), vol. 27 (1899). LC

> Continuation of the work by Sir Ernest Satow in vols. 7 and 9 of the Transactions. A number of gestures occur in the rituals described.

Football's sign language. Popular mechanics, vol. 88 (Sept. 1947), p. 114-115. Illust.

Football sign language. Look magazine (Oct. 11, 1938). Illustrated.

Förstemann, E. W. Commentary of the Maya manuscript in the royal public library of Dresden. Cambridge, Mass. 1906. Fla.

Fortescue, Adrian. The mass, a study of the Roman liturgy. New York. 1937. Fla.

———. The uniate eastern churches. The Byzantine rite in Italy, Sicily, Syria, and Egypt. Ed. by George D. Smith. London, 1923, xxiii, 244 p. Bibliog. p. xi-xxi.

———. The ceremonies of the Roman rites described . . . with plans and diagrams. London, 1918, xxxi, 441 p. Bibliography p. xxiii-xxviii; 8th ed. published at Westminster, Maryland, xvii, 431 p. Illustrated. Bibliog.

> Provides additional notes about U.S. practices.

———. The orthodox eastern church. London, Catholic truth society, 1916, xxxiii, 451 p. Bibliog.
> On orthodox rites with gestures, p. 418-427.

Fouché, P. Rapport du congrès international de linguistique de Rome de 1930. (Manuscript)
> Enlarges the concept of the word 'language' to include gesture communication. (Item from Tcheng-Ming, p. 2)

Franz, Adolph. Die kirchlichen Benediktionen im Mittelalter. Freiburg, 1909. LC

Frazer, Sir James G. Folk-lore in the Old Testament. New York, 1927.
> See chap. 9, p. 343-349, 'The silent widow.'

———. The golden bough. New York, 1941, abridged ed., vol. I, p. 240.
> On folding of hands & crossing legs as taboo.

French gestures. Speaking of pictures. Life magazine, vol. 21 (Sept. 16, 1946), p. 12-15. Illust.

Frenzen, Wilhelm. Klagelieder und Klagegebärden in der deutschen Dichtung des höfischen Mittelalters. Würzburg, 1938. 85 p. (Univ. of Michigan Library).

Frijda, Nico H. The understanding of facial expression of emotion. Acta psychologica, vol. 9 (1953), p. 294-362.
> The evidence of the existence of an unambiguous and invariable meaning in the different aspects of facial expression is offered. Films & photos were used in the experiment.

Frois-Wittmann, J. The judgment of facial expression. Journal of experimental psychology, vol. 13 (1930), p. 113-151.
> A statistical analysis of judgments of facial expressions of grief, sulking, pleasure, gloating, aversion, etc.
> See Hulin & Katz, The Frois-Wittmann pictures of facial expression, Journal of experimental psychology, vol. 18 (1935), p. 482-498.

Furlani, G. Short physiognomic treatise on the Syriac language. American oriental society journal, vol. 39 (1919), p. 289-294.

Fussel, Paul Jr. The gestic symbolism of T. S. Elliot. A journal of English literary history, XXII (1955), p. 194-211.
> The author studies Elliot's use of "details of the exterior workings of the human body."

Gaines, Jack. El lenguaje silencioso de la radio. La voz de los Estados Unidos, (Sept.-Oct. 1950), p. 2-4. With 12 illust. Published by the U. S. Department of State.
> On the gestures in use in radio broadcasting stations.

Gallaudet College for the Deaf.
> Established 1864, Washington, D. C. Supported by the U. S. Congress. An institution for higher learning for the deaf. An important source of accurate information of the sign languages.

Gaster, M. The exempla of the rabbis. London, 1924, p. 269, no. 443.
> On the Zeichendisput.

Geiger, Paul & Richard Weiss. Atlas de folklore suisse. Basel, 1951, 76 p. 16 maps. LC
> See distribution of greeting formulas.

Gent, J. B. Chirologia, or the natural language of the hand. Composed of the speaking motions and discoursing gestures thereof. London, 1644. Illustrated.

George, S. S. Gesture of affirmation among the Arabs. American journal of psychology, vol. 27 (July 1916), p. 320-323.
> Corrects the error of H. Petermann, in Reisen im Orient, 1860, I, p. 172, who stated that Arabs in affirmation shake their heads as we do in negation. The error needed correction, among other reasons, because it was broadcast in Wundt's celebrated Völkerpsychologie, 1904, vol. I, first part, p. 180.

Gesell, Arnold L. Studies in child development. New York, 1948.

Gestural language discussed. New York Times (Dec. 22, 1946), VI, 49:2.

Gestures betray you. Science digest (Dec. 1948), p. 42.

Gestures not inherited. Science news letter, vol. 40 (1941), p. 243.
> Denies Nazi theory that gestures are a part of each human being's racial inheritance.

Giese, Fritz. Psychologie der Arbeitshand. Berlin-Wien, 1928, viii, 325 p. Illust.
> A learned monograph on the hand with several items of value for gesture study. See, for example, p. 304 ff.

Giles, P. A far-travelled story. Aberdeen university review, vol. I (1913-1914), p. 259-264.
> Concerning the Zeichendisput: man holds up fingers, opponent misinterprets meaning.

Gilmore, Art & G. Y. Middleton. Television and radio announcing. Hollywood Radio Publishers, Inc. Hollywood, Calif., 1949, 3rd edition. 283 p.
>See chap. I for illustrated explanations of radio & television gestures used in broadcasting.

Giner de los Ríos, Gloria. Cumbres, New York, 1955.
>For miscellaneous illustrations of gestures from the Hispanic peninsula from Roman times on, see p. 58, 61, 62, 93, 99, 100, 105, 114, 157, 188, 191.

Ginneken, J. van. Een nieuwe Ontdekking der Taalwetenschap. Onze Taaltuin, 1938. Univ. of Chicago Library.

——— Principes de linguistique psychologique, essai de synthèse. Paris & Amsterdam, 1907. Quotes Wundt briefly on gesture, p. 530-531. LC

Giraudet, A. Mimique, physiognomie et gestes. Paris, 1895.
>Actors' manual. Often obscure in meaning.

Give-away gestures. Coronet magazine, vol. 30 (May, 1951), p. 116-119. Illust.
>On the gesture studies of Alfred E. Johns, who is called "a pioneer consulting psychologist," and who says there is "a definite relationship between our innermost feelings and the seemingly innocent gestures we make." Unqualified interpretation of ten illustrations of gestures. Caveat.

Glossary of international gestures. Travel, vol. 22 (Feb. 1914), p. 35.
>Author writes on need for a handbook of international gestures and describes several from far-off places.

Goldberg, Isaac. The wonder of words. New York, 1938, p. 53-57 et passim. LC
>Author quotes from Sir Richard Paget's theory of the gestural origin of language and gives strong support to it. Suggest that gestures may well have been the first grammar. A fundamental work in gesture study.

Goldstein, K. Journal of psychology, vol. II (1936), p. 301 ff.
>On the neurology of gestures.

Goldziher, I. Über Gebärden und Zeichensprache bei den Arabern. Zeitschrift für Völkerpsychologie und Sprachwissenschaft, vol. XVI, 4 Heft, Berlin, 1886, p. 369-386.

——— Zauberelemente im Islamischen Gebet. Orientalische Studien. Theodor Nöldeke zum 70. Geburtstag. I. Giessen, 1906, p. 303-329. LC
>See especially p. 320 ff., paragraph 5, on prayer gestures.

Gondal, I. L. Parlons ainsi de la voix et du Geste. Paris, new ed. of Gigord, 1912, p. 407-419.

Goodenough, F. L. & M. A. Tinker. Relative potency of facial expression and verbal description of stimulus in the judgment of emotions. Journal of comparative and physiological psychology, vol. 12, (Dec. 1931), p. 365-370.

Goodland, Roger. A bibliography of sex rites and customs. London, 1931, 752 p. LC

> Bibliog. p. 1-667. Index of subject matter, p. 669-750, with brief note on content of each item cited. A fundamental source of bibliography of sex gestures. See index under *gestures, evil eye, dances,* etc.

Gougaud, L. Dévotions et pratiques ascétiques du moyen âge. Collection "Pax," vol. XXI, Paris, 1925, p. 1-42.

> About prayer gestures.

——— La prière des bras en croix. Rassegna Gregoriana, vol. VII (1908), p. 345-46.

Graff, Paul. Geschichte der Auflösung der alten Gottesdienstlichen Formen in der evangelischen Kirche Deutschlands. Göttingen, 1937 & 1939. 2 vols. (1st ed. 1921). LC

> On gestures used in worship, see vol. I, p. 284-286; a brief but scholarly discussion. Each volume contains a lengthy bibliography and good index. Protestant viewpoint. See under Kreuzeszeichen of both vols. for material on crossing oneself. A fundamental work.

Grain market of the world. Chicago Board of Trade, Chicago, 1948, p. 7-8.

> Signs of grain-market traders.

Grajew, Felix. Untersuchungen über die Bedeutung der Gebärden in der griechischen Epik. Dissertation. Freiburg, 1934.

Grande encyclopédia portuguesa e brasileira. Lisboa & Rio de Janeiro, 1945.

> See vol. 12 (gesticulado) and vol. 15. (linguagem).

Gratiolet, Pierre. De la physionomie et des mouvements d'expression. Paris, 1865. 436 p.

> "L'étude de la physionomie, c'est-àdire des modifications que les sentiments, les sensations et les idées impriment à la forme d'un être vivant, a fixé des temps anciens l'attention des artistes, des poëtes et des philosophes . . . Les mouvements d'expression sont en effet les éléments du langage spontané de l'homme et des animaux. Il ne peut donc être indifférent de

rechercher quel lien secret unit les signes spontanément employès aux choses signifiées, c'est-a-dire a l'idée ou au sentiment qu'ils manifestent . . ." A precursor of the large tribe of psychologists interested in gesture.

Gray, Giles W. Index to the quarterly journal of speech, vols. I-XL (1915-1954). Dubuque, Iowa, 1956, 338 p.

> Speech teachers and elocutionists formerly placed gesture in the center of their curriculum. See, for example, Flora N. Kightlinger, The star speaker, Jersey City, 1894, p. 63-109. For more than a quarter of a century gestures have been made to play second fiddle. The Giles index bears this out: it cites few items on gesture. Under numerous headings, such as "speaking" and "speech," for example, one finds evidence of a vast ocean of repetition among Speech pedagogues, with minor attention paid to gesture.

—— Problems in the teaching of· gesture. Quarterly journal of speech education, vol. 10 (1934), p. 238-252.

> "Gesture is of importance . . . in a speech curriculum; . . . it contains enough matter to be included under the general heading of 'content;' . . . it can be, and is being, taught." The article impresses one adversely by its 'erudition' and lack of unity.

Green, B. P. A handbook in the manual alphabet and the sign-language of the American deaf. Ohio, 1916. (Item from Critchley)

Green, Lili. Einführung in das Wesen unserer Gesten und Bewegungen. Berlin, 1929. 152 p. 257 illustrations. LC

> On the psychology of movement & gesture.

Gregory, Joshua C. Magic, fascination, and suggestion. Folk-Lore, vol. 63 (1952), p. 143-151.

> The power of the eyes to "work spells."

Grief: see Apuleius, The golden ass, book IV (pulling one's own hair and beating chest). See also Yvaine (in old French lines 1155-1159; 1300; 1412-1413).

Griffith, Helen Stuart. Sign language of our faith. Baltimore & New York, ca. 1939. Illust.

> Greek religious gestures.

Grimm's household tales. See Paula Cohn.

Grimm, Jakob. Deutsche Rechtsaltertümer. Göttingen, 1854, xx, 970 p. Fla.

Grimm, Wilhelm. Ueber die Bedeutung der deutschen Fingernamen. Kleinere Schriften, herausgegeben von Gustav Hinrichs. Berlin, 1883, vol. 3.

Grinde, Nick. Handmade language. Saturday evening post, vol. 221 (July 10, 1948), p. 34-35 and 117-120. Illust. in color.

> Mainly about Indian sign language. Author suggests we take a tip from the Redskins and establish an international sign language of the world.

Grohne, Ernst. Gruss und Gebärden. Handbuch der deutschen Volkskunde; herausgegeben von W. Pessler. Potsdam, n.d., vol. I, p. 315-324.

Grolleau, Charles, and Guy Chastel. L'ordre de citeau; la Trappe. Paris, 1932.

> Chap. IV contains a section on silence and one page of explanation of a number of the gestures used for communication among the Trappist monks. "Ces signes ne sont pas un alphabet; le vocabulaire en est si restreint qu'il ne permet pas de tenir sur tout sujet une longue conversation, mais il est expressif, imagé et rapide."

Groschuf, G. Abhandlung von den Fingern, deren Verrichtungen und symbolische Bedeutung. Leipzig und Eisenach, 1756.

Groslier, Georges. Danseuses cambodgiennes, anciennes et modernes . . . préface de Charles Gravelle. Paris, 1913, 178 p. Illust. LC

> Dancing & gestures of Cambodia.

Gruenert, M. Das Gebet im Islam. Prague, 1911.

Gubbins, J. K. Some observations on the evil-eye in modern Greece. Folk-Lore (London), vol. LVII (Dec. 1946), p. 195-198.

> Some 'cures' for the evil eye in Greece. Fear of this curse is not restricted to rural areas.

Guilford, J. P. Experiment in learning to read facial expressions. Journal of abnormal psychology, vol. 24, p. 191-202. Bibliographical footnotes. Tables.

———— & M. Wilke. New model for the demonstration of facial expressions. American journal of psychology, vol. 42 (1930), 436-439. Bibliog. footnotes. Illust.

> Authors believe their new model is an improvement over previous ones.

Guenther, J. Kultur der Geste—Geste des Kultus. Gestalt, vol. 3 (1930-31), p. 41-48.

Guentert, H. Grundfragen der Sprachwissenschaft. Leipzig, 1925.

Gui, Bernard (early 14th century) in The Waldensian heretics, The portable medieval reader, New York, 1949, p. 215: oath taken by man touching the Holy Bible and raising hand.

Gurnee, H. Analysis of the perception of intelligence in the face. Journal of social psychology, vol. 5 (Feb. 1934), p. 82-90.

Hackett, Francis. Francis I, New York, 1935, p. 139-140.
>Describes coronation gestures.

Hacks, Charles. Le geste. Paris, 1887. 492 p. Illust. NYP
>Somewhat obscure speculation about meaning of gestures. Wordy. No notes or bibliography.

Haddon, A. C. The gesture language of the Eastern Islanders [of Torres Straits.] Cambridge anthrop. exp'd. to Torres Straits, Reports. Vol. 3 (1907), p. 261-2. NYP

Hadley, Lewis F. Indian sign talk. Chicago, 1893. LC
>Hadley's most extensive work. 268 octavo leaves preface; 192 p. of dictionary of gestures alphabetically arranged. Each page has three gestures figured and beside each the equivalent in English. Nearly 600 signs, about 800 illustrations; appendix. 53 p. of reading matter in signs, mainly holy writ. Hadley's Indian name was Ingonom Pashi, given him by the Indians.

——— A lesson in sign talk. Fort Smith, Arkansas, 1890. LC
>Designed to show the movement of the hands in the Indian gesture language. Twelve illustrations, some Scripture texts, including the Lord's prayer, rendered in drawn signs.

——— A list of the primary gestures in Indian sign-talk. Anadarko, Indian Territory, 1887. LC, NYP
>Very rare. Only 75 copies were published. According to Seton (q.v.), Hadley was a missionary who "made a study of Sign Language in order to furnish the Indians with a pictographic writing, based on diagrams of the signs, and meant to be read by all Indians, without regard to their speech. Pointing to the Chinese writing as a model and parallel, he made a Sign Language font of 4000 pictographic types for use in his projected work. He maintained that 110,793 Indians were at that time sign-talkers and be proposed to reach them by Sign-Language publications."

——— Wolf lame & the white man, by In-go-nom-pa-shi [pseud.] Fort Smith, Ark., 1890. 8 p. Illust. NYP
>Indian sign-language.

Haegy, Joseph & L. M. LeVavasseur. Manuel de liturgie et ceremonial. Paris, 1935. 2 vols. Illust. 16th ed.
<blockquote>This volume evolved from a work by G. Baldeschi.</blockquote>

Haiding, Karl. Von der Gebärdensprache der Märchenerzähler. Folklore Fellowship Communications, Helsinki, Academia Scientiarum Fennica, number 155 (1955), 16 p. 11 photos.
<blockquote>How gestures make Austrian folktale narrator's style more effective.</blockquote>

——— Kinderspiel und Volksüberlieferung. München, n.d.

Haigh, Arthur E. The Attic theatre. Oxford, 1907. 3rd ed.

——— The tragic drama of the Greeks. Oxford, 1896. viii, 499 p. Illust. Fla.

Hall, G. S. Gesture, mimesis, types of temperament, and movie pedagogy. Pedagogical seminary, vol. 28 (June, 1921), p. 171-201.

Handbuch des Taubstummenwesens. Osterwieck, 1929. P. 36-42; 321-334; 706 ff.

Handclapping. For applause, Plutarch, Lives, ed. Harvard classics, New York, 1910, vol. 12, p. 236 & 277; meaning disapproval, Job XXVII, 23; also J. B. Greenough and Geo. L. Kittredge, Words & their ways in English speech, New York, 1911, p. 223: "explodo meant, in Latin, to drive off an actor by clapping the hands. Tabu: Rousseau, Confessions, book VIII, says there was never any clapping at the theater when the king was present.

Hand (raising) Throughout the United States, in schools & in public meetings, one raises his hand to indicate that he has an answer, a request, or a question, for the teacher, or speaker, or chairman of the assembly.

Handshake. Homer, Odyssey, books I and XVII, 236. For statue with handshake, see H. H. Powers, The message of Greek art, Chautauqua, N.Y., 1913, p. 106, 113.

———. Preservation of. When Queen Elizabeth II of England visited Lagos, Nigeria, a Nigerian attemped to 'preserve' her handshake. After he shook hands with the Queen, "he clenched his fist tightly and kept it so. Then he carefully 'carried' the Queen's touch away through the throng and quietly pressed his palm to his forehead in a gesture of irrevocable ritual assimilation." Manchester Guardian Weekly (England), Feb. 9, 1956.

———. The handshake is the customary ritual used for closing the unprogrammed meeting of the Quakers (Society of Friends).

———. Sign in Cochabamba, Bolivia, police station (in 1945): "Por higiene sírvase no dar la mano." The handshake is ubiquitous and frequent there.

Hands joined in a pact. See Shakespeare, 3 Henry VI, IV, 6, lines 38-40.

Hanna, Jay. Sign talk speeds radio drama. Popular mechanics, vol. 63 (May 1935), p. 702-704; 120 A. Illust.

> An interview with Jay Hanna, who originated many of the gestures used in radio broadcasting stations, according to this article.

Hansen, Marian. Children's rhymes accompanied by gestures. Western folklore, vol. VII (1948), p. 50-53. Bibliog. p. 50.

Harbage, Alfred. Elizabethan acting. Publications of the modern language association of America, vol. LIV (1939), p. 685-708.

Harrison, Jane. Prolegomena to the study of Greek religion. Meridian books. New York, 1955, xxii, 682 p. Illust. Fla.

> Studies Greek religion primarily as seen through its ritual, through research in what Greeks *did* in addition to what they *thought*.

———. Themis; a study of the social origins of the Greek religion ... with an excursus on the ritual forms preserved in Greek tragedy by Gilbert Murray ... Cambridge, 1927, xxxii, 559 p. Illust. Fla.

Hartmann, J. Kley. Repertorium rituum. Paderborn, 1940.

Hartmann, Johann Ludwig. Pastorale evangelicum. Nuernberg, 1678.

> Chapter 14 gives a thorough discussion of prayer gestures. Some of the headings are de gestibus et actionibus, de vultu, de capite, de fronte, de superciliis, de oculis, de naribus, de buccis, de labiis, de collo, de cervice, de manibus, de digitis, de pedibus, de universo corpore. (—Paul Graff, Geschichte der Aufloesung ..., vol. I, p. 284, note 6)

Harwood, Mrs. Eliza J. How we train the body; the mechanics of pantomimic technique. Boston, 1933. Illust. 142 p. LC

> Gestures & gymnastics.

Hastings, James. A dictionary of the Bible. New York, 1908, et seq.
> Fundamental. See, for example, under "the annual touching of the temple" of Bel-Merodach by kings for purpose of establishing afresh their title, vol. I, p. 213; see passim under hand, bowing, kissing, embracing, etc.

———. Encyclopedia of religion and ethics. New York, 1912.
> Fundamental. See under bowing, hand, embrace, kiss, etc.

Hayes, F. C. Should we have a dictionary of gestures? Southern folklore quarterly, vol. IV (1940), p. 239-245.

———. Gesture. Encyclopedia americana. New York, 1941 ed. et seq. Illust.

———. Just a gesture. Collier's magazine, vol. 109 (Jan. 31, 1942), p. 14-15. Illust.

———. Gestos o ademanes folklóricos. Folklore Américas, vol. XI (1951), p. 15-21.

———. Beckoning. American notes & queries, vol. I (1941), p. 142.

Hayner, P. C. Expressive meaning. Journal of philosophy, vol. 53 (1956), p. 149-157. Bibliog. footnotes. Fla.
> An attempt to state in philosophical terms some of the main characteristics of 'expressive meaning." Author sees a fundamental difference between expressing and describing something. Gestures often express meaning, whereas words 'tell" meaning in a fundamentally different way. When we speak of joy, we 'describe' it. When we gesticulate for joy, we exhibit it, and thus give an expressive meaning.

Haywood, Charles. A bibliography of North American folklore & folksong. New York, 1951. xxx, 1292 p. Fla., LC

Heffner, Hubert C. et al. Modern theatre practice, a handbook for nonprofessionals. New York, 1946, 3rd ed.
> See bibliography of books on acting, p. 472-478.

Height (indicating with hand)
> If one indicates height of an animal, the palm is turned down; if of a child, the palm is vertical and the thumb thrust up. (Mexico, Guatemala)

Heiler, Friedrich. Die Körperhaltung beim Gebet. Orientalische Studien, dedicated to Fr. Hommel at his 60th birthday, Leipzig, 1918, vol. II, p. 168-177.

———. Das Gebet. Munich, 1920. Translated in English: Prayer, New York; 1932; in French: La Prière, Paris, 1931.
>See chap. VI on prayer gestures.

———. Gebet. Die Deligion in Geschichte und Gegenwart. Tübingen, vol. 2, 1927-31, p. 872-874.

Hekkenbach, Josephus. De nuditate sacra sacrisque vinculis. Giessen, 1911. (Item cited by Hoffmann-Kreyer)

Helwig, P. Die Seele als Aeusserung. Leipzig and Berlin, 1936. iv, 124 p. LC

Hemsterhuis, Franz. Vorlesungen über den Ausdruck der verschiedenen Leidenschaften durch die Gesichtszüge, übersetzt von Schaz. Berlin, 1793. LC

Herrgot. Vetus disciplina monastica. Paris, 1726.
>Item from Critchley.

Hersey, John. A single pebble. New York, 1956.
>Bowing in China; also bargaining with gestures of hands concealed under cloth: p. 66; 125-126; 40.

Hertz, R. La prééminance de la main droite. Mélange de la sociologie religieuse et folklore. Paris, 1928.

Hewes, Gordon W. The anthropology of posture. Scientific American, vol. 196, (Feb. 1957), p. 123-132.

———. World distribution of certain postural habits. American anthropologist, vol. 57, no. 2, part 1 (April 1955), p. 231-244.

———. The one-legged resting position. Man, vol. 53 (1953), art. no. 280, p. 180.

Higgins, Dan. D. How to talk to the deaf. Chicago, 1942. Illust. LC
>An excellent manual for anyone wishing to learn to speak to the deaf with gestures. The author has presented his material well, but occasionally makes somewhat overly enthusiastic claims for the universality and 'values' of gesture study.

Hildebrand. Die Sitte des Hutabnehmens. Germania, vol. 14, 1869, p. 125-128.

Hildburgh, W. L. Apotropaism in Greek vase-painting. Folk-Lore (London), vol. LVII (Dec. 1946), p. 154-178.
> See especially p. 154 on paintings made for protection against the evil-eye. Valuable bibliographical footnotes.

———. Psychology underlying the employment of amulets in Europe. Folk-Lore (London), vol. LXII (March 1951), p. 242 ff.
> Article includes brief comments on the sign of the fig & other gestures.

Hillway, Tyrus. Melville's use of two pseudo-sciences. Modern language notes, vol. 64 (March 1949), p. 145-150. Bibliographical footnotes.
> Brief commentary on Melville's use of physiognomy and phrenology.

Hissing & shaking fist. Zephaniah II, 15. (Goodspeed translation)

Hitler (Adolph) gesticulating during a speech. See "An ill wind was blowing," Look magazine (Oct. 1946). During the second world war numerous pictures were run in magazines showing the tribal gesture of the Nazis. The gesture was accompanied by the phrase, 'Heil Hitler!'

Hocart, A. M. The mechanism of the evil eye. Folk-Lore (London), vol. 49 (1939), p. 156-157.

Hoffman, Walter James. The beginnings of writing. New York, 1895. Illustrated with drawings from North American Indians, the ancient Egyptians, etc. Fla.
> See especially chap. VI, Gesture signs & attitudes.

Hoffmann-Krayer, Eduard & H. Bächtold-Stäubli. Handwörterbuch des deutschen Aberglaubens. Berlin & Leipzig, 1927 ff. LC
> See vol. 3, p. 327-338, under Gebärde. This is one of the fundamental works for gesture study. I have selected a number of the important bibliographical items for inclusion in the present bibliography. One should consult this work under Gebet, Kreuzzeichen, Feige, Eid, etc. Vol. 10 contains an exhaustive index.

Hofsinde, Robert. Indian sign language. New York, 1956, 96 p. Illust. NYP
> Indian sign-talk for boys.

———. Talk-without-talk. Natural history, vol. 47 (1941), p. 32-39. Illust. NYP

Hole, Christina. Popular modern ideas on folklore. Folk-Lore (London), vol. 51 (1955), p. 321-27.
>   See p. 327-329 on Witchcraft & the evil-eye.

———. Witchcraft in England. New York, 1947. 167 p. Illust. Bibliog. p. 161-163.

———. English folklore. London, 1940. viii, 183 p. Illust. Bibliog. p. 177-178.

Homage. Sub-Shieks in the Near East kiss the hand and dagger, the latter in a holster, of the Chief Shiek. (—Scene from "The Seven Wonders of the World"—a Cinerama)

Homer, Iliad. Pallas Athene nods a negative, VI, 3; Achilles does likewise, XXII, 205; Zeus makes affirmative & negative gestures, with head, XVI, 250; Zeus' negative shakes Olympus, I, 526-30
>   It was evidently same in ancient Greece as today: a toss of the head backwards, often accompanied by a click of the tongue. For further references on ancient gestures, see Ebeling's concordance of Homer.

Hoops, Jihannes. Reallexikon der germanischen Altertumskunde. Strassburg, 1911, 4 vols.

Hopkins, E. Washburn. The sniff-kiss in ancient India. Journal American oriental society, vol. 28 (1907), first half, p. 120-134.

Hoppe, Ruth. Die romanische Geste im Rolandslied. Koenigsberg, 1937. viii, 184 p. Literaturverzeichnis, p. 179-184. Col. Univ. Lib.
>   Thoroughly documented. Many quotations from Old French. See, for example, under Die Schwurgesten, Das Küssen, etc.

Horst, J. Proskynein. Zur Anbetung im Urchristentum nach ihrer religionsgeschichtlichen Eigenart. Neutestamentliche Forschungen. Dritte Reihe, 2. Heft. Gütersloh, 1932. LC

Horvorka, O. von, and A. Kronfield. Vergleichende Volksmedezin. Stuttgart, 1908.

Howes, H. W. Gallegan folklore III. **Folk-Lore (London), vol. 40** (1929), p. 53-61.
>   On the evil-eye, see p. 53-55 and 388-389.

Howitt, Alfred W. Native tribes of southeast Australia. London & New York, 1904, p. 723-35.
> General commentary on gesture language plus specific comments on the use of gestures by aboriginal Australians; also a brief listing and word descriptions of number of gestures in common use. No illustrations.

——— Proceedings of the Australian Association for the Advancement of Science, vol. 2 (1890-1891), p. 637-646. (Item from Critchley)

Huart, C. Khams(a). Encyclopedia of Islam. Leyden, 1908-29, vol. II, p. 897: The hand of Fatima.

Huber, E. Evolution of facial musculature and facial expression. Baltimore, 1931.

Hughes, Henry. Die Mimick des Menschen auf Grund voluntarischer psychologie. Frankfort, 1900, xi, 423 p. 119 illustrations. LC
> Descriptive and interpretive.

Hughes, Russell Meriwether (La Meri, pseud.). The gesture language of the Hindu dance. New York, 1941. xviii, 100 p.
> This work contains 236 illustrations, each one descriptive of this mimic language; also complete glossary and dictionary of Sanskrit terms involved as well as an alphabetic list of the hand poses.

Hulin, W. S. & Katz, D. Frois-Wittmann pictures of facial expression. Journal of experimental psychology, vol. 18 (Aug. 1935), p. 482-498.
> A variation to the approach of Frois-Wittmann in his experiments.

Idelson, Abraham Z. The ceremonies of Judaism. New York, ca. 1930. 14 p. Illust. Fla.

——— Jewish liturgy & its development. New York, 1932. xix, 404 p. Fla.

Indian sign. Life magazine (July. 1954), p. 33. Illustrated.
> Brief piece on Chou En-Lai's visit to India & his rapid adoption of the Indian salutation of bowing with hands together.

Irwin, F. W. Thresholds for the perception of difference in facial expression and its elements. American journal of psychology, vol. 44 (Jan. 1932), p. 1-17. Bibliographical footnotes. Tables.
> Author combines philosophy and statistics.

Islam, The world of. Life magazine, vol. 38, no. 19 (May 9, 1955), p. 72-92.
> Provides photos of some Islamic gestures.

Jackson, Hughlings. Brain, vol. I (1878), p. 304. Fla.
> On the neurology of gestures.

Jahn, Ulrich. Hexenwesen und Zauberei in Pommern. Breslau, 1886.
> See under Blick, p. 81 ff.

James, William. The principles of psychology. New York, 1950. 2 vols.
> See vol. II, chap. XXV, p. 442-485, the emotions; chap. XXVII, p. 594-616, hypnotism; chap. XXIII, the production of movement.

James, W. T. A study of the expression of bodily posture. Journal of genetic psychology, vol. VII (1932), p. 405-37.

Janet, Pierre. L'intelligence avant le langage. Paris, 1936.
> See p. 102-104; 202-211; 228. Extension of the meaning of the word language to include more than auditory speech: "On pense avec tout son corps." Others who have developed this theme are H. Delacroix, G. Dumas, J. Morlaas, A. Ombredane, M. P. Fouché, J. Vendryes.

Jarden, E. and W. Fernberger. Effect of suggestion on judgment of facial expression of emotion. American journal of psychology, vol. 37 (Oct. 1926), p. 565-570.
> Continuation of experimentation with the Piderit (q.v.) model for demonstrating facial expressions.

Jelgerhuis, Johannes. Theoretische lessen over de gesticulatie en mimiek. Amsterdam, 1827. 10 vols. Illust. NYP
> Gesture, expression, stage-costume.

Jenness, Arthur. Effects of coaching subjects in the recognition of facial expressions . . . Journal of general psychology, vol. 7 (July 1932), p. 163-178. Bibliog. p. 177.
> A repetition of F. H. Allport's (q.v.) experiment and a criticism of his conclusion.

——— Differences in the recognition of facial expression of emotion. Journal of general psychology, vol. 7 (1932), p. 192-196. Bibliog. Tables.
> Are men or women more adept at recognizing facial expressions? What facial gestures are most readily and accurately identified?

Jespersen, Otto. Language. Its nature, development and origin. New York, 1922. Fla.

Jewish encyclopedia, The universal. New York, 1943. 10 vols.
> Rich in illustrations, many of which show judaic gestures. See, for example, vol. IX, p. 170; vol. II, p. 167-170 (benedictions, Jewish *barach;* vol. I, p. 100, 202, 128, 333, 561-568; vol. IV, p. 561, 604, 606, 607, 234-236; passim.

Johannesson, Alexander. Origin of language. Reykjavik, 1949.
> "In Nature, Feb. 5, 1944, I published a summary of the researches contained in this book in which I tried to prove that the most important class of the 2200 Indo-European roots could be explained as imitation by the speech organs of the movements of the hands. I came to the conclusion that human speech consists of three main sources: 1) Spontaneous sounds of emotion, such as sneezing, coughing, sounds of joy, sorrow, anger, etc. 2) Imitations of sounds in nature e.g. the song of the birds, the whistling of the wind . . . etc. 3) sounds produced by the organs of speech as imitations of the movements of the hands. Of these three classes the last is the most important for the development of language."—P. 10-11. A fundamental work.

—— Gestural origin of language. Reykjavik & Oxford, 1952.
> Johannesson expands his original study of the gestural origin of language found in his "Origin of language." He presents evidence from six unrelated families of languages, namely, Indo-European, Hebrew, Archaic Chinese, Polynesian, Turkish, and Greenlandic. Cemal Enisoglu, of Turkey, says in the preface: "It can be said that, by his (Johannesson's) researches, the problem of the origin of languages has been solved in the main lines and that the investigation of all the other languages from this point of view will only add further evidence." Johannesson gives a bibliog. of his own studies, p. 17.

—— Origin of language. Nature, vol. 157 (1946), p. 847-848.

—— Origin of language (in imitation of gestures). Nature, vol. 162 (1948), p. 902.

Johnny Belinda (movie). In this Academy-award cinema of 1948, the deaf mute, played by Jane Wyman, repeats the Lord's Prayer in sign language.

Jones, M. R. Studies in nervous movements. Journal of general psychology, vol. 29 (1943), p. 47-62; 303-312. Bibliog. Tables
> Effects of mental arithmetic on the frequency and patterning of the nervous, or autistic and spontaneous, movements of a group of high-school boys.

Jones, Putnam Fennell. A concordance to the Historia Ecclesiastica of Bede. Cambridge, Mass., 1929. Fla.

Johnson, Ben. Volpone, III, 1, lines 20-24: Mosca describes a parasite's gestures.

Jorio, Andrea de. La mimica degli antichi investigata nel gestire. Naples, 1832. xxxvi, 380 p. Illust. Color plates. NYP, LC

> An indispensable volume. Drawn up in alphabetical order. Several good indices. Author says if we understand the contemporary (i.e., 1832) gestures of the Neapolitans, we may interpret the gestures of the ancients of Greece, especially those on Greek vases.

Jousse, Marcel. Etudes de psychologie linguistique: le style oral, rythmique et mnémotechnique chez les verbo-moteurs. Archives de philosophie, vol. 2, cahier 4(1923). Bibliog. p. 236-240. NYP

> Study of gestures from psychologist's point of view.

———— Le mimisme humain et l'anthropologie du langage. Revue anthropologique, nos. 7-9 (1936), p. 201-215.

> Jousse defends the thesis that "le son, phonomimiquement émis par le geste laryngo-buccal, ne vient donc d'abord que renforcer, préciser et parfaire audiblement la signification de tel ou tel geste manuel mimique et visible. Peu à peu, chaque geste manuel caractéristique ou transitoire est doublé d'un adjuvant sonore . . . On nous avait autrefois répété, au cours de nos études de Linguistique grammaticale, que les racines indo-européennes, sémitiques ou chinoises avaient toutes des significations concrètes. Nous en conaissons maintenant la raison . . ." Jousse's theorizing has been carried on by A. Johannesson, q.v., A. V. Thomas, q.v., Sir Richard Paget, and others.

———— Etudes sur la psychologie du geste. Les rabbis d'Israel. Les récitatifs rythmiques parallèles. Paris, 1929. LC

———— Méthodologie de la psychologie du geste. Revue des Cours et Conférences (1931), p. 201-218.

> Man is gesticulative because he is a mimic.

———— Du mimisme à la musique chez l'enfant. Paris, 1935.

———— La métaphor gestuelle chez l'enfant et le primitif. Cours inédit professé en 1941 a la Faculté de Médecine de Paris. The list of titles of these 20 lectures on gestures is found in A. V. Thomas, L'anthropologie du geste, Revue anthropologique (1941), p. 193. Here also is a bibliography of the works on gestures of M. Jousse published to date.

Kapsalis, Peter. Greek gestures. Doctoral dissertation. Johns Hopkins University, 1946. (Unpublished)

Kaulfers, W. V. Curiosities of colloquial gestures. Hispania, vol. XIV (1931), p. 249-264.

——— Handful of Spanish. Education, vol. 52 (Mar. 1932), p. 423-428.

Keller, Helen. Story of my life. New York, 1903.

——— The world I live in. New York, 1920.

Kelley, Samuel R. Tableaux d'art consisting of instantaneous changes of the emotions, for public entertainment. Boston, 1885. LC

Kemp, R. Mains de l'homme, agiles et fortes, implorantes ou dominatrices. France illustration, no. 373 (630-636a), Dec. 6, 1952.

Kesson, J. The cross & the dragon. London, 1854. LC
In chap. 18, description of several secret signs of the Grande Secret Society of China.

Kightlinger, Flora N. The star speaker. Jersey City, 1894. Illust. See 63-109; places gesture in central position in the curriculum of elocution.

Kinesics. Defined in the 1954 Addenda of Webster's International Dictionary: "The systematic study of body motions (as winks, shrugs, waves) in their relation to communication between persons."

King, Chas. William. The gnostics and their remains. London, 1864. xvi, 251 p. Illust. 2nd ed. 1887. LC, Col. Univ. Lib.
Ancient gestures passim; see, v.g., p. 120-123, "Recognition by means of symbols."

King, Wm. S. Hand gestures. Western folklore, vol. VIII (1949), p. 263-264.
10 gestures collected from Univ. of California students.

Kingson, W. K. & Rome Cowgill. Radio drama acting and production: a handbook. New York, 1950. 2nd ed. Illust.
See Studio sign language, p. 358-363.

Kirchner, Paul C.. Jüdisches Ceremoniell. Nurnberg, 1726. 226 p. 18 plates. Col. Univ. Lib.
Gestures passim; see, v.g., plates opposite p. 226, 172, 76; also chaps. on prayer.

Kiss (greeting): see Genesis XXXIII, 4; Odyssey, Book XVII (Wm Cullen Bryant transl.)

Kissing Bible. Required in many states of the United States before testimony can be legal. In Sept. 1941 it was announced over NBC, the "Light that shines in the darkness" program, that when a certain Dr. Francis Schiffler discovered 20 courts in New York City without Bibles, he supplied them. In the South some communities have two Bibles in court: one for whites and the other for Negroes.

Kjellén, Nicolaus A. W. De actione ad sacras orationes applicata . . . Upsaliae, 1801. 12 p. NYP

Klages, Ludwig. Ausdrucksbewegung und Gestaltungskraft. Grundlegung der Wissenschaft vom Ausdruck. Mit 41 Figuren. Leipzig, 1923, et seq. Duke Univ. library.

Klauser, Theodor and others. Reallexikon für Antike und Christentum. Stuttgart, 1954, vol. 2, p. 474-482: Böser Blick. Bibliog.
> On the evil eye in ancient times in the Orient, Egypt, Greece and Rome, among the Hebrews, and in Christendom.

Kleen, Tyra af. Mudrās. The ritual hand-poses of the Buddha priests and the Shiva priests of Bali, with an introduction by A. J. D. Campbell . . . with 60 full-page drawings by the author. 1924. LC
> Excellent drawings and explanation of the significance of ritual hand gestures, or Mudrā. Miss Tyra de Kleen was Swedish artist and traveller. (See also Kulturen der Erde, vol. XV, 1923).

———— Hand-poses of the priests of Bali. Asia, vol. 24 (Feb. 1924), p. 129-131. Illust.

Kleinpaul, Rudolf. Das Leben der Sprache und ihre Weltstellung. Leipzig, 1888-93. See vol. I: Sprache ohne Worte; also under Mienen und Gebärden, p. 158-177; 277-388; rump signs, p. 271. ff.

Kline, L. W. and D. E. Johannsen. Comparative role of the face and of the face-body-hands as aids in identifying emotions. Journal of abnormal psychology, vol. 29 (Jan. 1935), p. 415-26.

Klineberg, Otto. Racial differences in speed and accuracy. Journal of abnormal and social psychology, vol. XXII (1927), p. 273-277.

Klitgaard, C. Skaelsord og foragtelig gestus. (Gestures of insult & contempt). Danske Studier, Copenhagen, 1934, p. 88-89.

Kneeling. Daniel VI, 10; Istiah XLV, 23; Luke XXII, 41 and many other places in the Bible. See also not of H. G. Bohn ed. of Samuel

Butler's Hudibras, London, 1907, p. 4, note 3: the Presbyterians of 17th century England refused to kneel at the Sacrament of the Lord's Supper & insisted on receiving it sitting or standing.

Knight, R. P. The symbolical language of ancient art and mythology. New York, 1892, p. 29-30.
> On the sign of the fig.

Kohlbrugge, Jacob H. F. Tier-und Menschenantlitz als Abwehrzauber. Bonn, 1926. 94 p. 182 illustrations. Bibliog. LC
> A carefully documented study. Numerous facial and some manual gestures.

Kortholt, Chr. De sacris publicis, debita cum reverentia praestantisque numinis metu colendis diatribe ascetica. Kiel, 1693
> See section on prayer gestures.

Krappe, A. H. Balor with the evil eye. New York, 1927. viii, 228 p.

—— The science of folklore. London, 1930.
> See p. 219; 283; passim. (Might have contained a chapter on Folkspeech, with sub-section on Folkgesture.)

Kraus, F. S. Slavische Volksforschungen. Abhandlungen über Glauben, Gewohnheiten, Sitten, Bräuche und die Guslarenlieder der Südslaven. Leipzig, 1908.
> See p. 170 on primitive gestures.

Krout, M. H. Autistic gestures. Psychological monographs, vol. XLVI (1935), p. 1-126.

—— Introduction to social psychology. New York, 1942. Illust.
> Gestures & speech, p. 322-323; deaf & dumb, p. 313-314; empirical gestures, p. 316; symbolism of gestures, p. 313-316. Similar gesture patterns of different groups have emerged from similar human experiences, p. 314-315. Gestures possess high utility, hence high survival value, p. 316. Theory of gestural origin of language, p. 322-328. See important bibliography of Chap. VI on symbolism.

—— Symbolic gestures in clinical study of personality. Transactions of the Illinois state academy of science, vol. XXIV (1931), p. 519-523.

—— The social & psychological significance of gestures. The journal of genetic psychology, vol. XLVII (1935), p. 385-412.

—— A preliminary note on some obscure symbolic muscular responses of diagnostic value in the study of normal subjects. American journal of psychology, vol. XI (1931), p. 29-71.

―――― Further studies in the relation of personality and gesture: a nosological analysis of autistic gestures. Journal of experimental psychology, vol. 20, March, 1937, p. 279-287.

―――― Understanding human gestures. The scientific monthly, vol. XLIX (1939), p. 167-172.

Krukenberg, H. Der Gesichtsausdruck des menschen. Stuttgart, 1913.

Kuekulhaus, H. Urzal und Gebärde. Grundzuege eines kommenden Massbewusstseins. Berlin, 1934. 248 p. Illust. LC

> Bild und Gebärde, p. 12-16.

Ku Klux Klan initiation.

> New members kneel and the Exalted Cyclops touches them on the shoulder with a sword, saying: "I dedicate you to the holy services of our country, our Klan, our homes, and each other and humanity. My Terrors and Klansmen, let us pray." Then all bow in prayer.

Kuenssberg, Eberhard, Freiherr von. Rechtsbrauch und Kinderspiel. Heidelberg, 1920.

―――― Schwurgebärde und Schwurfingerdeutung. Die Rechtswahrzeichen. Heft 4. Freiburg, 1941; also in Zeitschrift für Schweizerische Recht, vol. 61, p. 384-420.

> Symbolism in law.

Kurze Abhandlung von der Händesprache, insoweit deren Merkmaale bei den alten Schriftstellern sich äussern, mit deren eigenen Beweistümern bestätiget. Cassel, 1750.

> Short treatise on gesture based on ancients.

Kuvalayanda. Asanas, Popular Yoga, London, 1938, vol. 1, part 1. LC

> See chap. on Yogi posture.

Labarre, Weston. The cultural basis of emotions & gestures. Journal of personality, vol. XVI (1947), p. 49-68.

> Surveys a variety of gestures from different parts of the world and concludes that they might offer new insights into psychology, psychiatry, and linguistics. Concludes also that gestures, like any form of social behavior, must be learned.

La Fin. Sermo mirabilis, or the silent language. London, 1692.

> An early advocate of a 'silent language of the fingers' for deaf mutes. Still earlier advocates were Jerome Cardano (1501?1575), and Pedro Ponce de Leon (1520-1584).

Lake, Evelyn F. C. Some notes on the evil eye round the Mediterranean basin. Folk-Lore (London), vol. 44(1933), p. 93-98.
> Not only humans, but horses, donkeys and cattle must be protected from the evil eye. Various protective measures are cited.

Lambert, Louis C. Mimetic expression: a study of gesture. The Expression Co., n.d., n.p.

La Meri. The gesture language of the Hindu dance. New York, Columbia University Press. n.d. 236 illustrations. Foreword by Ananda K. Coomaraswamy.
> Alphabetical list of Hasta-Mudras (hand poses). Glossary of Sanscrit terms used.

—— Spanish dancing. New York, 1948. xii, 186 p. Photographs, appendix & bibliography.
> See especially the braceo, arm postures & gestures.

Landis, Carney. Studies in emotional reaction: II. Journal of comparative psychology, vol. 4 (1924), p. 447-509. Bibliog.
> Landis outlined with charcoal on the subject's face the regions of the principal muscles, then subjected him to various stimuli: popular music, the Bible, pornographic pictures, firecrackers, etc. During the response a photo of the facial expression was made. Article I of this series appeared in the Jour. of experimental psychology, Oct., 1924. Article II contains a brief historical account of methods of experimentation used in gesture study, with brief estimate of each. See also R. C. Davis.

—— and Wm. A. Hunt. The startle pattern. New York, ca. 1939. Fla.

Landsberger, Franz. A history of Jewish art. Cincinnati, 1946. 369 p. Illust.
> See striking 'alms box', p. 35, with outstretched hand; 'carrying the law', p. 36; hands portrayed on the Mizrach Tablets, p. 41.

Langdon, S. Gesture in Sumerian & Babylonian prayer. Journal of the royal Asiatic society (1919), p. 531-556.

Lange, Fr. Die Sprache des menschlichen Antlitzes. Eine Wissenschaftliche Psysiognomik und ihre praktische Verwertung im Leben und in der Kunst. Munich & Berlin, 1937; 2nd ed. 1939. 228 p. Illust. LC
> A scientific physiognomy and its practical application in life and art.

Langer, Susanne K. Feeling and form—a theory of art (Developed from Philosophy in a New Key, New York, 1942 et seq.) New York, 1953. 431 p. Bibliog.

> Gesture defined, p. 180; in dance, 174-187, 196, 203-204; as symbolic form, 175.

——— Philosophy in a new key. A study in the symbolism of reason, rite, and art. 6th ed. New York, 1954.

> See index for references to gesture.

Language of gestures. Folk-Lore (London), vol. XXXIX (1918), p. 312.

La Piere, R. T. and P. R. Farnsworth. Social psychology. New York, 1949, 3rd ed. xiii, 626 p. Bibliography and index, p. 549-619.

> See p. 69-79 on gestures.

Lasaulx, E. de. Die Gebete der Griechen und Römer. Würzburg, 1842.

> See appendix for listing of lectures.

Lasch, Richard. Der Eid. Seine Entstehung und Beziehung zu Glaube und Brauch der Naturvölker. Stuttgart, 1908.

La Trappe in England. Anon. London, 1937.

> Item from Critchley.

Lawrence, Robert Means. The magic of the horseshoe with other folklore notes. Boston, 1898. 344 p.

> A number of gestures are described passim; see, for example, 'The folklore of common salt.'

Lawson, Joan. Mime. London, 1957. 208 p. Illust.

> Historical basis of gesture, with extensive materials on practical and technical aspects.

Laying-on-of-hands. "Maud A. Snow, music teacher, testified that as a result of laying on of hands [of evangelist Chas. S. Price], she had been cured of tumors, Bright's Disease, and other ailments."— Manitoba Free Press, as quoted in H. L. Mencken, Americana, New York, 1926, p. 240. For cure of sick cow by similar process, see Americana for 1925, p. 20. For laying-on-of-hands in the Methodist ritual, see photo of the censecration of a bishop, Florida Times-Union (July 1, 1952).

Lazarillo de Tormes. Anon., 16th cent. Spain.
> See chap. III: hidalgo who left his home in Castile to go to live elsewhere in penury in order to avoid doffing his hat to another local gentleman *first*.

Le Bidois, M. G. De l'action dans la tragédie de Racine. Paris, 1900.
> See chap. IV for the role of the human face in the drama of Racine.

Legg, Wickham J. English church life from the restoration to the tractarian movement. London, New York, 1914. xvi, 428 p.
> Gestures: p. 168-172, etc. See esp. chap. VI. provides numerous notes and references.

Leibnitz, G. G. Opera omnia. Geneva, 1768. Vol. VI, part 2, p. 207.
> Item from Critchley.

Leite de Vasconcellos Pereira de Mello, Jose. A figa. Estudo de etnografia comparativa, precedido de algunas palavras a respeito do 'sobrenatural' na medicina popular portuguesa. Porto, 1925.

―――― A linguagem dos gestos. Ethnografia artistica, III. Separata da Alma Nova, Lisboa, vol. II. (1917) Illust. LC

―――― Sur les amulettes portugaises. Lisbonne, 1892. 12 p. LC
> Comments on the figa amulet and its uses.

Lekis, Lisa. The origin and development of ethnic Caribbean dance and music. Doctoral dissertation in manuscript. University of Florida, 1956. 282 p. Bibliography.
> Indispensable for Latin American dance and other gestures.

―――― Ethnic and folk dances of Latin America. Unpublished ms. prepared for the University of Florida Press, 1956. Bibliog. and maps.
> Indispensable for study of Latin American gestures, especially those of the dance.

Lemoine, Jean-Gabriel. Les anciens procédés de calcul sur les doigts en Orient et en Occident. Revue des études islamiques, 1932, p. 1-60. Illust. LC
> Carefully documented study of the three Oriental systems of finger-counting, their antiquity, where in near and far east finger-counting was and is used by traders (who cover their hands with a cloth during a trade), the relationship between finger-counting and the origin of our number system, etc. Excellent bibliography of studies bearing on dactylonomy and its relationship to the question of the origin of numbers, p. 56-57. Of particular interest to mathematicians.

Leonhard, Karl. Ausdruckssprache der Seele, Darstellung der Mimik, Gestik und Phonik des Menschen. Berlin, 1949. 507 p. Illustrated. LC

Lepper, J. W. Famous secret societies. London, n.d.

Leroy, D. La raison primitive. Paris, 1927.

Lersch, P. Die Bedeutung der mimischen Ausdruckserscheinungen für die Beurteilung der Personlichkeit. Industrielle Psychotechnik, vol. V (1928), p. 178-183. LC

> Interpretation from the psychologist's point of view, with a view to practical use in industry.

——— Gesicht und Seele. Grundlinien einer mimischen Diagnostik. Dresden, 1932. 167 p. Illust.

Lessing, Th. and W. Rink. Der Körper als Ausdruck. Schriftenreihe zur Gestaltkunde. Leipzig, n.d.

Lévy-Bruhl, L. Les fonctions mentales dans les sociétés inferieures. Paris, 1918. 3rd ed. Translated as Primitive Mentality, New York, 1923. Fla.

Lewis, George. A series of groups, illustrating the physiognomy, manners and character of the people of France and Germany. London, 1823. 52 engraved plates.

Lexicon of trade jargon. Federal writers' project. Ms. in the Library of Congress (Washington)

> Some reports on the sign language in use in various American trades. This item was called to my attention by H. L. Mencken in a personal letter in which he said he planned a section on gestures in the forthcoming supplement to The American Language (1948). This section never materialized.

Lhomme, Jean. La loma. La revue mondiale, 1920.

> "Essai d'un langage par geste dans la vie et au théâtre."—Jan Doat, L'expression corporelle, p. 71.

Liebrecht, Felix. Zur Volkskunde. Alte und neue Aufsätze. Heilbronn, 1879. xvi, 522 p. LC

Lifer, Serge. Annotation of movement, kinetography (In Russian). Art Publishing House, Moscow, 1940.

Lifting cap (for respect). See Arthur Tilley, Literature of the French renaissance, Cambridge, I, p. 281. Sixteenth century —professors' way of honoring Cujas and Turnèbe —in Germany . Some British dons today lift their caps each time they mention Jesus' name.

Linguistic structure of native America. New York, 1946. 423 p. Fla.
> On American-Indian languages. Fla.

Lippert, Julius. Kulturgeschichte der Menschheit in ihrem organischen Aufbau. Stuttgart, 1886-7. 2 vols. (Translated by G. P. Murdock as The evolution of culture, New York, 1931) Fla.
> Vol. I, p. 150 ff. and 159 ff. of German ed., on primitive gestures.

Liturgy, see Ritual.

Lodge, Oliver. Džamutra, or the bridegroom. Folk-Lore, vol. 46 (1935), p. 244-267, 306-330.
> Marriage customs from southern Serbia. A number of gestures passim; vg, p. 314. Index p. 408.

Lommatzsch, Erhard. System der Gebärden, dargestellt auf Grund der mittelalterlichen Lit. Frkrs. Dissertation, 1910 in dem Sammelband Philosophische Dissertationem 13, Berlin, 1910.
> Only preface and Chap. D were published.

——— Deiktische Elemente im Altfranzösichen. 1 Teil in Hauptfragen der Romanistik, Heidelberg, 1922. 2 Teil in Jahrbuch für Philologie, 1 Bd., München, 1925. LC
> Concerns the hinweisen, or pointing-out element in Old French, with numerous examples cited.

——— Gebärden, Verhandlungen der Berliner Gesellschaft für Anthropologie, Ethnologie und Urgeschichte. 1890, p. 329 ff.

Long, J. Schuyler. The sign language. A manual of signs, being a descriptive vocabulary of signs used by the deaf of the U.S. and Canada. 2nd ed. Des Moines, Iowa. 1918. Profusely Illust.
> Author was principal of the Iowa School for the Deaf. Good index. Brief study of 'The sign language,' p. 5-11.

Long, Major Stephen H.. The Indian language of signs. Expedition to the Rocky Mts., vol. I, 1832, p. 378-394.
> "Gives 104 signs. The earliest extensive vocabulary on record."—E. T. Seton, Sign talk, p. vi.

Loomis, C. Grant. Sign language of truck drivers. Western folklore, vol. V (July 1956), p. 205-206.
> Several gestures with meanings.

Love, Kenneth. Egyptian land reform.. New York Times (Oct. 23, 1955) Sec. I, p. 9..
> A *fellah* speaks about his gestures and liberty.

Löw, J. Die Finger in Literatur und Folklore der Juden, Gedenkbuch zur Erinnerung an D. Kaufmann. Breslau, 1900.

Lucretius, T. De rerum natura, Berkeley, California, 1917. Book I.
> See discussion of origin of language.

Lutz, Florence. The technique of pantomime. Berkeley, California, 1927. 174 p.
> 510 emotions or situations illustrated: admiration, belief, depression, prayer, etc. 140 pages of tables.

——— Inheritance of the manner of clasping hands. American naturalist, vol. 42 (1908), p. 195-196.

Lutz, H. F. Speech consciousness among Egyptians and Babylonians Osiris, vol. 2 (1936), p. 1-27. LC
> See esp. the descriptions and explanations of certain ancient Egyptian and Babylonian gestures, p. 14-16.

Lyall, A. Italian sign language. 20th Century, vol. 159 (June 1956), p. 600-604.

Lynn, J. G. & D. R. Smile & hand dominance in relation to basic modes of adaptation. Journal of abnormal & social psychology, vol. 38 (1943), p. 250-276.

Mac Gowan, D. I. List of signs in Historical magazine. vol. X (1866) p. 86-97. Fla.

Madinier, Gabriel. Conscience et Mouvement, Paris, 1938. ix, 481 p. Thesis., University of Paris. Bibliog. p. 463-469. LC
> The "psychology of movement."

Magazines, Gestures portrayed in American. See Life, Look, Pic and other picture magazines, especially the issues published during World War II, for pictures of gestures, articles about them, front covers based on them, etc. See (for example, Baptismal gesture, Look (March 10, 1942); It's a patriotic gesture now (i.e., thumbing a ride) (Life, July 6, 1942); front-cover gesture of "silence", pleading for secrecy in troop movements, Collier's (April 11, 1942); numerous photos of Fascist salute in Germany, Italy, and Spain, etc.

Malecot, J. L. "A note on gesture and language," Quarterly journal of Speech, vol 13, p. 439.

> Generalities on gesticulation; contrasts American and French gestures; says "in French the organs of speech are in position before the emission of sound; that is, the vocal gesture precedes the sound . . . In English the vocal gesture is simultaneous with the emission of sound . . ." Claims proper gesticulation in American fashion improves his pronunciation of English.

Mallery, Garrick. Sign language among North American Indians compared with that among other peoples and deafmutes. U.S. Bureau of Ethnology, I (1879-80), p. 263-552. Profusely illustrated. LC

> Introduction is general study of gestures with speculation on the origin of speech, brief paragraphs on gestures in the lower animals, in children, in deaf mutes, of the blind, etc. This is an elaborate and valuable compilation from many conrtibutors other than North American Indians. See p. 600 for index of gestures; p. 265 for list of 268 illustrations. A fundamental work.

―――― A collection of gesture-signs and signals of the North Amercan Indians with some comparisons. Miscellaneous Publications, Bureau of American Ethnology, 1880. No. 1, 329 p. (Only 250 copies printed).

――――Greetings by gesture. New York, 1891. Reprinted from the Popular science monthly, Feb.-Mar., 1891. LC

―――― Introduction to the study of sign language among the North American Indians. Washington, D.C., 1880. iv, 72 p. Illust. LC.

―――― Israelite and Indian; a parallel in plane of culture. New York, 1889. 47 p. Reprinted from Popular science monthly (Nov. Dec. 1889). LC

―――― The gesture speech of man . . . Salem, 1881. 33 p. Reprinted from vol. 30 of Proceedings of the American association for the advancement of science. LC.

> Indians of North America; sign language; salutations.

Malott, D. W. Grain and its marketing. Grain Exchange Institute, Chicago, 1940. Chap. 19, p. 6, brief explanation of the meaning of system of trading gestures in use in the grain market. No illustrations.

Mangin, H. Die Hand, ein Sinnbild des Menschen. German translation, Zürich, 1951, 229 p. Illust. LC French translation: La main, portrait de l'homme, Paris, 1947, 283 p. Illust. LC

A treatise on hand-reading by a man whose vocation, according to the preface, is the human hand; "il est constamment penché sur des mains." The frontispiece lists numerous other works on the hand by Mangin. His name for his lucubrations is "la chiroscopie (qu'il ne faut pas confondre avec la chiromancie.)"

Mantegazza, Paolo. La physionomie et l'expression des sentiments. Paris, 1885 (Eng. trans. Physiognomy and Expression. London, 1904.) Fla.

> Constantly refers to Darwin's book on human emotion. Mantegazza's approach is often dispersive. Considers emotional gestures of all sorts, those with and those without conscious meaning. Some of Mantegazza's ideas seem a little precipitously made, as for example classification of all gestures, in which those of South American Indians are *glum,* those of the Hottentots *stupid,* of the Maoris *ferocious,* of the Europeans, *intelligent!*

——— Physiologie de la Douleur. Paris, 1888, 350 p. See Troisième Partie, "Expression de la Douleur."

> Author often seems overly-speculative in his opinions; mainly concerned with autistic (automatic, nervous) gesticulation and expression.

Mantzius, Karl. A History of theatrical art. London, 1903. 10 vols. LC See vol. 1, p. 233-234. Gesture (Pantomimus) became an independent art in Roman times. "During the Empire pantomime became the favorite branch of the drama, and nearly supplanted the spoken drama." p. 233.

Marañón, Gregario. The psychology of gesture, Journal of nervous and mental diseases, CXII (1950), p. 469-497. Translated from the Spanish: Psicología del gesto, Havana, 1937, 86 p.

> Somewhat vague generalities about gestures; v.g. el gesto y el cine, el gesto del domador (Hitler, Mussolini, Stalin), etc. Abstracted in Psychological abstracts, Vol. 25 (1951), item no. 4535.

Marinus, Albert. Langage et Manuelage, II Tesaur, vol. 4 (July-Dec., 1952). A study of the relationship of gestures to language.

Marr, N. Ja. See Thomas, Lawrence L.

Martín de Castañega, R. P. Fray. Tratado de las supersticiones Y Hechicerías. Madrid, Sociedad de Bibliófilos Españoles, 1946, vol. 17, chap. 14: Que el aojar es cosa natural y no hechicería." Originally published in 1529 in Logroño.

> Who makes use of the evil eye, how it functions, who are most adept (old women), who is most susceptible (babies); methods of protection. This volume was officially endorsed by Bishop Alonso de Castilla, who ordered its distribution among the clergy in order to discredit certain widespread superstitions.

Mas-Latrie. Louis de. Dactylogieo Dictionnaire de Paleographie. Paris 1854. vol. 47. p. 179-366, An extended study of finger-talking as used by the deaf, the svages, etc. About 30 American Indian signs are described and compared with those of the deaf. No illustrations."—Seton, Sign Talks, p. xi.

Masson-Oursel, P. "Le rôle des attitudes dans la conception indienne de la vie." Psychologie et vie, vol. IV (1930), p. 23-24.

Masters, John. Bugles and a Tiger, Atlantic Monthly (Dec. 1955), p. 38. The GURKHA (of India) "gesticulates little . . . and points with his chin, not his hand."

Mather, Eleanor P. Barclay in brief. Pendle Hill pamphlet. Wallingford, Pennsylvania, 1948. 72 p.
> Contains selections from a 1676 edition of Barclay, including chapter on Hat and Knee, p. 52-54. Quakers herein described were persecuted for refusing to kneel, bow or uncover heads "to any but God."

Mawer, Irene. The art of mime, London, 1932. With 32 illustrations and bibliography. LC
> The author is esthetically and educationally concerned with all bodily movements of actors. Chapter II is entitled: "Gesture and the gesture apparatus. Hands, wrist, elbows and shoulders." The language is occasionally obscure.

Maya Indians. Annual report of the Regents of the Smithsonian Institution for 1941, Smithsonian Publication 3651. Washington, D.C., Govt. Printing Office, 1942, 583 p. General appendix contains: Benj. Lee Whorf, "Decipherment of the Linguistic Portion of the Maya Hieroglyphs," p. 479-502.

——— Russian explains hieroglyphic find. New York Times, (Oct. 12, 1952), p. 22. Brief article on the Maya studies of Y. V. Knorozov, Soviet ethnologist, who hopes to decipher Maya glyphs. Much is made of gestures as a key to the enigma.

McCoy, Colonel Tim. Rough Rider . . . and Authority on the Indian Sign Language, Harper's Bazaar (June, 1936), p. 79; p. 141-144. Forty terms "well known to Westerners" defined.

Mead, George H. Mind, self and society. Chicago, 1934. Bibliography. Fla.
> See 'conversation of gestures', and "gestures" in index.

Mead, Margaret and McGregor, Frances. Growth and culture, New York, 1951.
   A photographic study of Balinese children. Some gestures.

'Meaningless' gestures may tell what you really think, Anon. Lincoln Evening Journal and Nebraska State Journal (July 4, 1955), p. 10.
   On Maurice Krout and his studies of autistic gestures.

Meggitt, Mervyn. Sign language among the Walbiri of Central Australia. Oceania, vol. 25, nos. 1-2, 1954, p. 2-16. Bibliog. p. 16. Fla.

Mehl, O. J. Das liturgische Verhalten. Göttingen, 1927.

Mehta, S. S. Modes of Salutation. Journal of the anthropological society of Bombay, vol. X (1914) p. 263-272.

Mehta, Ved. A donkey in a world of horses. Atlantic monthly, vol. 200 (1957), p. 24-30.
   Effects on a young blind boy from Indian when he hears a husband and wife kiss in public in America (p. 25); his reaction to the American handshake (p. 25); he thumbs his first ride (p. 30).

Meier, Frederick. Character reading for fun and popularity. Philadelphia, 1945. viii, 344 p. Illust. Permabooks. (Phrenology, hand, gesture). Thoroughly unreliable.

——— Character reading made easy. New York, Permabooks, 1949. viii, 183 p. Illust. (Chap. VII is about gestures.) Thoroughly unreliable.

Mencken, H. L. Treatise on the gods. New York, 1946. Prayer gestures, p. 117 ff.

Merriam, Alan P. The Hand Game of the Flathead Indians. Journal of American folklore, vol. 68, (1955), p. 313-324.
   An Indian gambling game which required elaborate pantomime and gesticulation.

Merryman, Montgomery. Portuguese: A portrait of the language of Brazil. Rio de Janeiro, 1945.
   See chapter on "The Eloquence of Brazilian Hands."

Meschke, Kurt. "Gebärde", Handwörterbuch des deutschen Aberglaubens. Berlin, 1927-42, 10 vols. See III, cols. 329-337.

Mexico Says it With Gestures, Pemex Travel Club Bulletin, III (Nov. Dec. 1941), p. 5-6. A popular illustrated article describing some 25 Mexican gestures.

Meyer, D. R. and others. Incentive, anxiety, and the human blink rate. Journal of experimental psychology, vol. 45, (March 1953), p. 183-187. 1 table. Bibliography.

> Increased incentive for the performance of a visually guided task will increase the blink rate. (Blinking becomes a gesture when it is clearly an expression of an inner emotion or tension.)

Michaels, J. W. A handbook of the sign language of the deaf. Atlanta, 1923. NYP

> (Especially for theological students.)

Michel, Karl. Die Sprache des Körpers. Leipzig, 1910. xlvi, 167 p. LC

> Several hundred photographs, with explanations of Rumpfgebärden, Ellbogengebärden; Aufzucken, Auffahren, Hände geballt, Zusammenzucken, Hände gespreizt, Fussgebärden, Hände an und auf den Kopf, an die Stirn, an die Schläfen, an Wangen, Nase, Mund, Kinn, Hals, etc. Illustrates many folk gestures. Valuable for actors and speakers. There is a 21 page introductory study divided into brief chapters; v.g., III, Ursprung der Gebärden; IV, Gebärden- und Wortsprache; IX, Körpersprache and Theaterschule.

Middle finger: called in Latin, Digitus Impudicus (also Infamis, Famosus and Verpus); in Italian, Dito Impuro.

> The middle finger thrust up with the other fingers lowered is one of the oldest and most widespread gestures known.

Migne, Jacques Paul, editor. Patrologiae. Paris, 1850, Tomus XC. LC

> Col. 685 et seq. "De loquela per gestum digitorum et temporum ratione". According to RW. Dom Gregory Dix, O.S.B., this essay is often included among collected works of the Venerable Bede but without justification. See the Venerable Bede, De Computo vel loquela digitorum, vol. 90, col. 295-298; also Sittl, q.v.

Minervini, G. Monumenti antichi inedite. Napoli, 1852.

> Hand gestures on Greek vases, including "sign of the fig."

Misiak, H. and Franghiadi, G. J. Thumb and personality. Journal of general psychology, vol. 48 (April 1953), p. 241-244. Bibliography.

> Brief history of the study of the hand from psychologist's point of view, plus brief account of a statistical study of the thumb and personality traits.

Modi, J. J. Tibetan salutations and a few thoughts suggested by them. Journal of the anthropological society of Bombay, vol. X (1914), p. 165-178.

Monde (Le) Plein de fols. Amsterdam, 1715. (Publ. also later as I Callotto Resucitato).
> See gesticulation portrayed in engravings of grotesque figures.

Monier-Williams, M. Buddhism, London, 1889. LC
> See passim on ritual poses considered among the requisites for union with the Supreme Spirit of the Cosmos.

Montaigne. Essays. Vol. II, chap. xii, et passim, on bowing, kissing, etc.

Montenovesi, Ottorino. Il linguaggio dei sordomuti in una pergamena veronese del 1472, Italy, vol. 2, 1933. p. 217-222. NYP

Moor, Edward. The Hindu Pantheon, London, 1810. 466 p.. plus 106 plates. LC
> Numerous gesticulating gods, some with extra heads and arms and hands. Copious notes by the compiler, but some gestures left unexplained.

Moor, A. P. Plates Illustrating the Hindu Pantheon. London, 1861.

Moreno Villa, José. Cornucopia de México. México, 1952. 149 p. 6 illust.
> See chap. 6: brief philosophical comment on the six gestures employed in Mexico for time, money, space, height of human being, height of animal, height of an inanimate object.

Moret, Alexandre. Le rituel du culte divin journalier en Egypte, d'apres les papyrus de Berlin et les textes du Temple de Séti Ier à Abydos. Musée Guimet (annales), vol. 14 (1902), p. 1-288. See index, p. 277-281, under embrassement, face, prosternements.

Morlaas, J. Du Mimage au Langage. L'Encéphale, no. 3 (1935), p. 197-208.
> Enlarging conception of word language.

Morley, Sylvanius J. The Ancient Maya. Stanford University Press, 2nd. ed., 1947, p. xxxii+520. Illust.
> See chapter on "Hieroglyphic writing, arithmetic and astronomy," p. 259-311. The Maya writing, which is ideographic, was still in use at the time of the Spanish conquest. It is today possible to read about one-third of the inscriptions. Most Maya manuscripts were burned by the Spanish clergy. Query: Since gestures abound in the Maya glyphs, would a study of cur-

rent gestures in Maya territory help to clarify the rest of the glyphs undeciphered? See, for example, the prominence of the gesticulating hand for FIRST in fig. 12, p. 232.

―――― An Introduction to the study of the Maya hieroglyphs. Smithsonian Bulletin no. 57. Washington, D. C., 1915.
> This work now superseded by later studies of Morley et al. On significance of "hand" to express idea of zero, see p. 102 et passim.

Mornet, D. French thought in the 18th century. New York, 1929, p. 52.
> The Chevalier de la Barre was decapitated (July 1, 1766) on two counts, one of which was failing to doff his hat as a religioss procession passed by.

Morris, Charles W. Foundations of the theory of signs. Chicago, 1938.

――――Signs, language & behavior. New York, 1946. 310 p.
> See especially preface and also large bibliography of works on sign language by estheticians, psychologists, linguists, and others. The author's 'scientific' style is on occasion obscure.

Moser, Oskar. Zur Geschichte und Kenntniss der volkstümlichen Gebärden. Sonderdruck aus Carinthia I, Mitteilungen des Geschichtsvereines für Kärnten, 144. Jahrgang, Heft 1-3, Klagenfurt, 1954, p. 735-774.
> Searching through the protocols of the county courts of Carinthia, the years 1570-1670, Moser partially reconstructs the daily life. He provides some illustrations and many descriptions of court gestures, ear-gestures (Ohrfeige), gestures of scorn, insult, greeting, with special emphasis on the sign of the fig which was used both for protection from the evil eye and as an insult. Numerous bibliographical items and footnotes. Index.

Moses' uplifted hands. Jehovah-Nissi (Jehovah my banner), the name given by Moses to the altar erected upon the hill where he sat with uplifted hands during the successful battle against the Amalekites, described in Exodus XVII, 15. Samuel Fellows, Bible encyclopedia, Chicago, 1907.

Mosher, Joseph A. Complete course in public speaking. New York, 1931. xxv, 631 p. Part II, Gesture, p. 1-82.
> Alarmingly exhaustive in its explanations, definitions, classifications, suggested techniques, table of hand-forms, high-, middle-, and low-planes, and other obscurities. LC

Moslem greeting. In greeting you, a Moslem may shake hands gently with you, or he may kiss your hand and raise his fingers to his lips afterward. "Do not laugh at him; it is his way of showing politeness."—From U.S. handbook for soldiers in North Africa. Quoted in Life magazine (Jan. 11, 1943), p. 60.

Moslem prayer gestures. Life magazine (Sept. 1946), p.. 47-50. Illust.

Mota, Fernando. El lenguaje secreto de los 'sin-patria.' Por esos mundos, Madrid (1915, año 16), p. 239-243. NYP
> Secret sign language.

Moulton, Robert H. Trading by signs. Leslie's weekly, vol. 122 (1916), p. 816-817. Illust. NYP

Mountford, C. P. Gesture language of the Ngada tribe of Warburton ranges, Western Australia. Oceania, vol. IX (1938), no. 2, p. 152-155. Fla.

Mühle, Günther W. and Albert Wellek. Ausdruck, Darstellung, Gestaltung. Studium generale, 5. Jahrgang, Heft 2, 1952, p. 110-130. LC

Müller, F. Max. The sacred books of the East. Oxford, 1885. (Vol. 28 of "The texts of Confucianism.")
> See p. 68 ff; also items 7,8,16,25,28,32, et seq. See passim other volumes of this series.

Müller, Günter. Ueber die geographische Verbreitung einiger Gebärden im östlichen Mittelmeergebiet und dem nahen Osten. Zeitschrift für Ethnologie, vol. 71 (1939), p. 99 ff.

Murray, H. A. Effect of fear upon estimates of the maliciousness of other personalities. Journal of social psychology, vol. 4 (Aug. 1933), p. 310-329. Bibliog. Tables.
> One's emotional reaction to facial expression colors one's judgment of others. (Murray used only five subjects in this experiment.)

Murray, Margaret Alice. The witch-cult in western Europe. Oxford, 1921. 303 p. Bibliography p. 281-285. Fla.

Murrin, R. Just a gesture, but what impression does it make? Good housekeeping, vol. 118 (Feb. 1944), p. 110-111. Illust.
> Very brief, somewhat undercautious, claims about the secret meaning behind nervous gestures.

Naidu, P. S. Hastas (being a study of the elementary hand poses in ancient Hindu dancing). New Indian antiquary, vol. I (1938), p. 345-361. Bombay. Illust. NYP

Nandikesvara: see Coomaraswamy.

National geographic magazine, vol. 90, no. 4 (Oct. 1946), p. 413-417.
> On Hispanic-American gestures.

Nazi salute. Numerous photos of this 'tribal gesture' were printed in American and foreign magazines and newspapers before and during World War II. The Communist salute also was widely used during the 1930's and 1940's and will be found in numerous picture magazines.

Neal, A. J. and M. S. Fry. The language of the silent world. Cardiff, n.d.

Nehemiah VIII, 6-7. Prayer gestures.

Nettesheim: see Agrippa von Nettesheim.

Neubert, F. Die volkstümlichen Anschauungen über Physionomik in Frankreich bis zum Ausgang des MAs. Dissertation, Munich, Erlangen, 1910. Duke University.

Oath: Sworn before altar, I Kings VIII, 31; hand used in oath-taking, Ezekiel XVII, 18; oath sworn by Jehovah, Genesis XXVI; 3.

Nod. Iliad, Book I, 526-530. Zeus' nod shakes Olympus. See also John XIII, 24 and Acts XXIV, 10.

O'Brien, Rev. John. A history of the mass and its ceremonies in the Eastern and Western Church. 14th ed., revised, New York, ca. 1879. xxii, 414 p. Illust. Bibliog. p. 393-395.
> See especially p. 181 ff. (the sign of the cross); p. 363-363, the kiss of peace; p. 381, kneeling and or dancing before the blessed sacrament; passim.

Ocean of story. Transl. by C. H. Tawney. London, 1924-28. 10 vols. Bibliography, vol. 9, p.. 171-335. Index, vol. 10.
> See vol. 1, p. 112; vol. 3, p. 150; also p. 37; vol. 9, p. 162; etc. This work is a fundamental source for determining approx. number of centuries some gestures have endured.

Oehl, Wilhelm. Fangen-Finger-Fünf, Collectanea Friburgensia. Freiburg, Switzerland: Veröffentlichungen der Universität Freiburg, XXXI (1933), p. 1-247; preface: N. S. Fasc. XXII.
> "This treatise . . . presents an exposition of the manner in which movements of the hand and certain phases in those movements have contributed to the build-up of a large number of word families showing wide ramifica-

tions of meaning. A vast mass of material is cited, much of it drawn from languages lying without the Indo-European fold."—Clarence Paschall.

Oesterley, W. O. E. The sacred dance. A study in comparative folklore. Cambridge, 1923, 234 p. Fla.

O'Hearn, Mick. Interpreter for deaf and dumb used sign language before she learned to speak. Gainesville (Florida) Daily Sun (Aug. 5, 1956).

> Story of Mrs. George Corrick, who learned to speak sign language before learning to talk. Both her parents were deaf and dumb.

Ohlert, Konrad. Rätsel und Rätselspiele d. Alten Griechen. Berlin, 1912.

> Some account of gestures passim.

Ohm, Thomas. Jesus und die Gebetsgebärde. Benediktinische Monatsschrift, 1947.

——— Die Gebetsgebärden der Völker und das Christentum. 1948. xvi, 472 p. 20 plates and 34 figures.

——— Die Anpassung an die Art und das Brauchtum der Nichtchristen in der Gebetsgestik. Neue Zeitschrift für Missionswissenschaft. (Switzerland), 1947, Heft 4.

Ombredane, André. Le langage, gesticulation significative mimique et conventionelle. Nouveau Traité de Psychologie, ed. by G. Dumas, Paris, 1933, vol. 3, p. 363-458.

> Extends the concept of the word language.

——— Études de psychologie medicale. II. Geste et action. Rio de Janeiro, 1944.

Orienter, A. Der seelische Ausdruck i.d. altdeutschen Malerei. Munich, 1921.

Orr, James. The international standard Bible encyclopedia. Chicago, 1937. 5 vols.

> See under hand, kneeling, bowing, kissing, praying, etc.

Ortiz Fernández, Fernando. Los bailes y el teatro de los negros de Cuba. Habana, 1951. Fla.

Osgood, Chas. E. and A. W. Heyer, Jr. Objective studies in meaning. II. The validity of posed facial expressions as gestural signs in interpersonal communications. American psychologist, vol. V (1940), p. 298. (abstract)

Osgood, Chas. O. A concordance to the poems of Edmund Spenser. Washington, 1915.
> See under embrace, kiss, etc.

Ostrup, J. Orientalische Höflichkeit. Formen und Formeln im Islam. Leipzig, 1929.
> On prayer gestures, see p. 25-42.

Ott, Edward Amherst. How to gesture. New York, 1902. x, 126 p. Revised and illustrated. LC
> Gestures used in elocution.

Otterstein, A. W. The baton in motion. Carl Fischer Co., New York. Illustrated.

Ovid. The metamorphoses. An English version of A. E. Watts, with etchings by Pablo Picasso. Berkeley, California, 1954.
> See p. 1 (phallic sign); p. 8 (nod); p. 11-12 (covering head; p. 12 (stones tossed behind one's back; p. 34 (beating one's breast); p. 34 (rending one's hair), et passim.

Pack, R. Catullus, Carmen V: abacus or finger-counting? American journal of philology, vol. 77 (Jan. 1956), p. 47-51. Bibliographical footnotes.
> Ingenious brief commentary on the counting gestures used as byplay in the poem of Catullus beginning: Vivamus, mea Lesbia, atque amemus ...

Paget, Sir Richard A.S. Gesture language. Nature, vol. 139 (Jan. 30, 1937) p. 198.

——— Gesture as a constant factor in linguistics. Nature, Vol. 158 (July 6, 1946), p. 29.

——— Human speech, New York & London, 1930. xiv, 360 p. Includes tables, charts, diagrams, illustrations. LC, Fla.
> A fundamental work for gesture study.

——— Sign language as a form of speech. Paper read at the Royal Institute of Great Britain (Dec. 13, 1935).

——— Babel, or the past, present, and future of human speech. London, 1930.
> See especially Chap. I: speculation on gestures as antedating language as a means of communication.

———— This English. London, 1935. xii, 118 p. Bibliographical footnotes. NYP

> See discussion on analogy between gesture and vocal speech. Also remarks of R. R. Marett in the preface, who comments on Paget's ability to "divine" the correct interpretation of 50% of a number of words in languages unknown to him by a study of the mouth gestures needed to form the words.

———— Origin of language, gesture theory. Science, vol. 99 (Jan. 7, 1944), p. 14-15.

> Paget debates E. L. Thorndike's theory that speech arose from "babbling." Description of how gesture theory was put to test at Oxford in 1929, when Pagett was challenged to "divine the correct interpretation" of a number of words in an unknown language by a study of their originating mouth gestures; he was over 50% accurate. Pure luck could have yielded only one in 100.

Palcos, Alberto. La vida emotiva. Buenos Aires, 1927. 225 p. LC

> A semi-popular study. Where do gestures originate in the body?

Pandeya, Gayanacharya A.. The art of Kathakali. Kitabistan, Allahabad, 1943. 163 p. Illust. LC

> Kathakali is the name of the ancient Indian folk-dance drama of Kerala. See especially chapters V thru VIII: symbolism of dance and temple gestures.

Pantomime. There is a voluminous literature on pantomime, most of which is omitted in this bibliography. Suggested histories of pantomime are Anton G. Bragaglia, Evoluzione del mimo, Milan, 1930, 393 p.; R. J. Broadbent, A history of pantomime, London, 1901, 226 p.; Albert Edward Wilson, King Panto: the story of pantomime, New York, 1935, 262 p., illustrated; bibliography p. 253 (Later ed., London, 1949); Alfred Auerbach, Mimik . . . Berlin, 1922.

Pardoe, T. E. Pantomimes for stage and study. New York, 1931, 393 p.

> Sample chapters: A brief history of pantomime; the body as an expressive agent. Sometimes obscure.

———— Language of the body. Quarterly journal of speech education, vol. 9 (June 1923), p. 252-258.

> A somewhat incoherently written plea for more emphasis on gestures in speech teaching. The language of gestures is "the one language the good Lord did not confuse at Babel," says the author.

Parrish, W. M. Speaking in public. New York, 1947.

> See pages 197-201 for advice on gesture and posture. Strong against the Delsartean system of gestures once in vogue.

Parrot, A. Documenta et Monumenta Orientis Antiqui. Paris(?), 1947-55. 5 vols. Vol. IV: Studia Mariana, 1950, xii, 138 p., 6 pl., 40 fig.

See "Cérémonie de la main at réinvestiture."

Parsons, E. C. Peguche, A study of Andean Indians. Chicago, 1945. 225 p. Illust. Index. Fla.

Paschall, Clarence. The semasiology of words derived from Indo-European *NEM. University of California publications in linguistics, vol. I (1943-48), p. 1-9. Bibliographical footnotes.

An able, thought-provoking plea for the serious inclusion of gesture in linguistic research.

Paschius, G. Inventa Novantiqua. Leipzig, 1700.

See p. 612 ff.

Paula, A. F. von. Paulys Real-Encyclopädie der klassischen Altertumswissenschaft. Stuttgart, 1894, 44 vols. Illust. Supplement, Stuttgart, 1903. Fla.

Pauli, Johannes. Schimpf und Ernst, herausgeben von Johannes Bolts. Berlin, 1924, 2 vols, vol. 2, p. 264-265. LC

Peale, O. F. Sign language of deaf mutes. American mercury, vol. 26 (Aug. 1932), p. 457-460. Fla.

Brief but revealing article. Deaf-mutes generally think in sign-pictures rather than in word-pictures. In sign language the niceties of grammar do not exist. Sign language is no potential Esperanto. Several common signs are described and explained. Author was formerly a teacher in a school for the deaf and dumb.

Pearn, B. R. Review of English studies, vol. XI (1935), p. 385-405.

Provides a full discussion of the Elizabethan "dumbshow."

Peck, Anne M. The pageant of middle American history. New York, 1947.

See p. 10-11 on Maya glyphs.

Pei, Mario. The study of language. Philadelphia, 1949.

See p. 9-17.

——— Gesture language. Life magazine, Jan. 9, 1950, p. 79-81.

A journalistic, illustrated article. Pei estimates that there are some '700,000 gestures tucked away in different parts of the world.'

——— All about language. New York, 1954.

See p. 18-21.

Pellegrim, A. Locution populaire: T'r'passras . . . come le t'chin Caal. Le Folklore Brabançon, vol. 3, p. 83. NYP

Pengniez, P. Cinématique de la main; la main du prestidigitateur. Presse Medicale, vol. 35 (1927), p. 123-125.

Perkins, J. B. France under the regency. New York, 1892. Fla.
> While momentous questions of finance, agriculture, and trade went begging, the dukes and the Parlement contended over the question of whether a duke or a president should first take off his hat. (-p. 242).

Perrot, Georges and Charles Chipiez. Histoire de l'art dans l'antiquité. Paris, 1882-1914. 10 vols. Illust. LC
> See esp. vol. I, "L'Egypte".

Peters. Gebet. F. X. Kraus Real Enzyklopädie der christlichen Altertümer. Freiburg in Breisgau, 1882.
> On prayer postures, see vol. I, p. 557-560.

Peterson, Fred. Of human laughter. Mexico this month, vol. III (1957), number 8.
> Brief commentary on the ancient smiling and laughing carved mask-like faces found near Vera Cruz. Illustrated.

Petrov, P. M. Children's gestures in experiment. Pedalogia, vol. II, (1931), p. 16-71.

Phillips, George L. Toss a kiss to the sweep for luck. Journal of American folklore, vol. 64 (1951), p. 191-196. Illust.
> Lift your hat or toss a kiss to a sweep, or have the bride kissed at the wedding by a sweep, and you will have good luck (England).

Phillot, D. C. A note on the mercantile sign language of India. Royal Asiatic society of Bengal, journal and proceedings. N. S., vol 2, (Calcutta, 1906), p. 333-334. NYP

——— A note on the sign-, code-, and secret-language, etc., amongst the Persians. Royal Asiatic society of Bengal, Journal and Proceedings. N. S. vol. 3 (Calcutta, 1907), p. 619-622. NYP

Picard, Max. Die Grenzen der Physiognomik. Erlenbach-Zürich und Leipzig. 1937, 191 p. Illust. LC
> A somewhat mystical-philosophical interpretation of physiognomy, with comments illustrated by masterworks of painting of faces; some are of celebrated people out of the past. The style is often eye-catching. Both image and gesture are emphasized and interpreted throughout.

Piderit, Th. Mimik und Physiognomik, 1867, 2nd ed. Wissenschaftliches System der Mimik und Physiognomik, Detmold, 1884.
> Suggests a demonstration model be constructed to reproduce facial expressions by an interchange of a number of mouths, eyes, brows and noses. (See also Boring and Titchener; Buzby, Jarden and Fernberger.)

Pitrè, Giuseppe. Gesti ed insegne del popolo siciliano. Rivista di letteratura popolare, I (1877), p. 32-43. Col. Univ. Lib.
> Brief, miscellaneous comments on several Sicilan gestures.

——— Usi e costumi del popolo Siciliano. Palermo, 1889, vol. II, p. 349 ff. On Sicilian gestures.

Plato, The Dialogues of. Translated into English by B. Jowett. Oxford, 1953. 4 vols. Index, vol. 4, p. 555-657. See under kisses, etc.

Plautus, The Rudens. Touch statue of Venus to sanctify an oath. (See K. Guinagh and A. P. Dorjahn, "Latin Literature in Translation". New York, 1947. P. 38: Plautus, The Rudens.)

Pliny (Plinius Secundus, C.) Naturalis Historia (On gestures used in connection with cunnus, see 28, 7, 23.)

Political Gesture—Peru. "Saluda con el brazo izquierdo en alto al jefe del partido como homenaje a la función que ejerce."—Harry Kantor, Ideología y programa del movimiento Aprista, Mexico, 1955, p. 220.

Pollenz, Philippa. Methods for the comparative study of the dance. American anthropologist, vol. 51, no. 3, p. 428-435.
> Suggests notational method to study and record dances (including gestures). Brief history of past notational methods. Short bibliography.

Pollice Verso (Thumbs down). H. P. Jones, Dictionary of Foreign Phrases and Classical Quotations. Edinburgh, 1925, p. 93.

Polti, Georges. Notation des Gestes. A. Savine ed. Paris, 1892.

Pontificale Romanum. Venetiis, 1835.
> Describes hand positions of ritual in minute detail.

Potter, Chas. F. Gesture, Standard dictionary of folklore. Funk and Wagnalls, New York, 1949.

Poucel, V. Mystique de la terre. Plaidoyer pour le corps. Paris, 1937. 308 p. LC
> Contents: Symbolique des formes-liturgie des fonctions.

Powell, Wilfred. Wandering in a wild country. London, 1883.
"An account of a three year residence in New Britain, with 14 good figures, showing digital origin of numbers, p. 254-264"—Seton.

Price, Willard. The Emperor next door. Harper's (July, 1941) p. 119.
On rigid rules for bowing to Japanese Emperor.

Praetorius. De Pollice. Leipzig, 1677, p. 101.
"Much information can be obtained from this book"—Elworthy, Horns, p. 154.

Prause, K. Deutsche Grussformeln in neuhochdeutscher Zeit. Wort und Brauch, vol. 19 (1930). LC

Preyer, W. Die Seele des Kindes. Leipzig, 1882. Fla.

Priestly Blessings (Hebrew). The universal Jewish encyclopedia. New York, 1948, vol. 8, p. 634.
Hands of priest were raised aloft; very ancient (nesi'ath kappayim); later the spreading out of the hands became characteristic symbol and is still carved on tombstones. For picture of this gesture, see F. Landsberg. A history of Jewish art, p. 41; other liturgy, see Elbogen, Der jüdische Gottesdienst, 1931, 2nd ed.

Prinz, Hugo. Altorientalische Symbolik. Berlin, 1915. On "thumb in mouth", see p. 20, Plate IV, Fig. 3.

Pritzwald, K. Stegmann von. Der Sinn einiger Grussformeln im Lichte kulturhistorischer Parallelen. Wörter und Sachen, vol. X. LC

Promies, Wolfgang. Mimik und Gebärde in Lope de Vegas "El Caballero del Milagro." Maske und Kothurn. Vierteljahrsschrift für Theaterwissenschaft. 3. Jahrgang, 1957. Heft 2. p. 116-27. (University of Vienna) Bibliographical footnotes.

Proskouriakoff, Tatiana. A study of classical Maya sculpture. Carnegie Institute of Washington. Publication No. 593 (1950), p. 23-31 et passim, on Mayan gestures of the glyphs.

Prudentius, see Roy J. Deferrari.

Psychological Abstracts, published monthly by the American Psychological Association.
See under gesture, face, hand, emotion.
Numerous experiments based on the meaning of gesticulation have been made and reported on, especially since about 1920. A selected number of these are included in the present bibliography. Students of folk gestures will find a mine of valuable information in the Psychological Abstracts.

Puglisi. M. La preghiera. Torino, 1928, p. 102-113 (about prayer gestures).

Putting Our Gestures in a Dictionary. Washington, D. C. Times Herald, (Sept. 19, 1943). Brief journalistic piece, illustrated with a cartoon.

Quackenbos, H. M. Archetype postures. The psychiatric quarterly, vol. 19 (Oct. 1945), p. 589-591.

Quedenfeldt, M. Verständigung durch Zeichen und das Gebärdenspiel bei den Marokkanern. Zeitschrift für Ethnologie. Band 22. Berlin, 1890.

Quiet hands spell a marriage; A deaf couple wed. Life, vol. 37 (Nov. 29, 1954), p. 56 ff.
> Very brief illustrated article on the marriage in London of a deaf Australian bride and a deaf English bridegroom. The gestures for "love", "church", etc. are photographed. According to the article, the English deaf uses both hands for the alphabet, the American uses only one.

Quintilian, M. F.   Institutes of Oratory. Book XI, chap. III. (cited by Critchley).

Rabanales, Ambrosio.   La Somatolalia. Boletín de Filología. (Universidad de Chile) VIII (1954-55), p. 355-378.
> General treatise on gestures, with classification of the different types of movements: expressive, communicative, descriptive, indicative, symbolic, active.

Rabelais, François. Gargantua, Book II. chap. 18 and 19: the "great scholar of England" who disputed by signs only with Panurge, without speaking, because the topics were so "abstruse, hard and arduous" that "words proceeding from the mouth of man will never be sufficient for unfolding of them" to the scholar's liking. See also Book I, Chap. 33, et passim.

Radio studio sign language.   See almost any radio manual; for example W. K. Kingson and Rome Cowgill, Radio drama acting and production: a handbook. New York, 1950.
> Studio sign language, p. 358-363. Illust.

Railroad signals (manual).   Illustration of these signals in Northern Pacific Railroad advertisement, Saturday Evening Post (August 8, 1953), p. 42. See also Pic Magazine (Dec. 23, 1941), p. 38-41.

Rands, Robert L. Comparative notes on the hand—eye and related motifs. American antiquity, vol. XXII (Jan. 1957), p. 247-257. Illustrated. Fla.

> These motifs are found in Southeastern USA, in Middle America, and on the Northwest coast of the U.S. Bibliography.

Ransom, Jay Ellis. Aleut Semaphore Signals. American Anthropologist, vol. 43, No. 3. pt. 1 (New Series) (July-Sept. 1941), p. 422-477. Illustrated.

Reallexicon.., Paul Merker und W. Stammler, Berlin, 1925-31. 4 vols.

> See under Gebärde.

Reed, E. A. Outline of physical expression. n.d. (University collection). Northwestern University Library.

Reich, F. Natürliche—Künstliche Gebärde? Blätter für Taubstummenbildung, vol. 37 (1924), p. 54-57.

Reinach, S. L'histoire des gestes. Revue archéologique (Paris) vol. XX (1924), p. 64-79.

> Gestures in early art of the ancients. Generalities. Reinach finds gestures in many statues, etc. down thru the ages too conventional. Says artists have failed to make fullest use of vast number of gestures.

Religious gestures: see studies of the rituals and ceremonies of any of the following: Hinduism, Zoroastrianism, Taoism, Confucianism, Jainism, Buddhism, Judaism, Christianity (any of various sects), Mohammedanism, Shinto, Mormonism.

Requeno Y Vives, Vincenzo. Scoperta della Chironomia, ossia, Dell'arte di gestire con le mani.., Parma, 1797. 141 p. NYP.

> 3 plates illustrating hand gestures for the Greek alphabet. Author maintains that it is necessary to learn the "arte di gestire con le mani, tanto necessaria per l'intelligenza de' Greci, e de' Romani Scrittori, e per la perfezione della moderne Pantomima ... La trascuratezza degli Eruditi è assai degna di riprensione in questa parte." Part I: Art of counting with the hands. Part II: Art of gesticulating with the hands necessary" pel resorgimento della greca pantomima."

Reuschert, E. Die Gebärdensprache der Taubstummen und die Ausdrucksbewegungen der Vollsinnigen. Leipzig, 1909.

Révész, G. Ursprung und Vorgeschichte der Sprache. Bern, 1946. 280 p. Bibliog. p. 267-271.

> Reviews theories of the origin of language, evaluates each one, proposes his own theory. See especially section on priority of gestural over vocal

language in time. Reviewed in Amer. journ. philol., vol. 669 (Oct. 1948); p. 425-430.

Reyes, Alfonso. La experiencia literaria. Buenos Aires, 1942.
> Discusses gestures in the first chapter; also p. 148.

Reynault, Félix. Le langage par geste. La nature, vol. XXVI (Oct. 15, 1898, 2e sémestre, no. 1524), p. 315-317.

Reynolds, Quentin. This is Sicily. Collier's Magazine, (Oct. 2, 1943). Illust.
> The Fascist salute fades away.

Ribsskog, Oyvin. Hemmelige Språk Og Tegn. Taterspråk, Tivolifolkenes språk, Forbryterspråk, Gatteguttspråk, Bankespråk, Tegn, Vinkelog Punktskrift. Oslo, 1945. 143 p. Illust.
> Secret language & signs. The communication devices used by various types, such as Gipsies, carnival people, burglars, street urchins and others.

Riemscheider-Hoerner, M. D. Der Wandel der Gebärde in der Kunst. Frankfurt, 1939.

Riffer, Vladimir. [Title in Greek] Indogermanische Forschungen, vol. 30 (1912) Strassburg.

Rijnberk, Gérard. Le langage par signes chez les moines. Amsterdam, 1953. 163 p. NYP
> Contains long alphabetical list of monastic gestures. Text in Latin with introd. and bibliog. in French. Thoroughly documented. Introd. essay on silence, its widespread requirements among monks from, at latest, the 4th century on, its use to avoid interruption of meditation of others, its meaning; necessity of sign language as a substitute for speech; early use among the ancients, esp. in the theater. Claims that in medieval Europe gestures were an international language for travelling monks. "Pour le but du présent article il suffit d'avoir démontré qu'il existait pendant l'Empire Romain un système de signes, SIGNA PRO LOQUELA, connus de la grande masse du peuple." (p. 8)

Rijnhart, Susie C. With the Tibetans in tent & temple. New York, 1901.

Rite: see Ritual.

Ritter, K. B. Das liturgische Gebet. III. Die liturgische Gebärdensprache. Christentum und Leben, vol. 11 (1936), p. 373-376.

Ritual. All ritual (and liturgy) is interspersed with gestures. The literature of ritualism is extensive. See, for example, the select bibliography of Eliza M. Butler, Ritual magic, Cambridge, England, 1949, p. 318-322. See also ritual books of the numerous sects.

Rochas d'Aiglun, Albert de. La mimique: Enseignée par l'hypnotisme. La Nature, Année 27, sem. 2, p. 252-254. (Paris, 1899). NYP
> Gestures & hypnotism.

——— Les sentiments, la musique et le geste. Grenoble, 1900. Illust. LC
> Expression, gesture, emotions, physical and psychological effects of music.

Rodríquez Arancibia, E. La Cueca chilena. Santiago, Chile. 1950. 42 p. LC
> Dance gestures.

Rolland, Eugène. Les Gestes, I. Mélusine, vol. III (1886-7), cols. 116-119. LC
> A number of gestures described from Turkey and France.

Rosa, Leone Augusto. Espressione e Mimica. Milan, 1929. NYP
> Provides over 300 sketches of gesticulating Italians by an observant artist.

Rose, H. A. The language of gesture. Folk-Lore, vol. 30 (1919) p. 312-315.

Rosekrans, R. L. Do gestures speak louder than words? Collier's magazine, vol. 135 (Mar. 4, 1955), p. 56-57. Illust.
> The gesture studies of Ray Birdwhistell, who is trying "to discover and decode the relationship between body motion and communication." He devised a system of shorthand "kinesics" for writing down all gestures, "everything from the tilt of the head to the crossing of the legs." Rather brittle interpretations given for secret meanings of nervous gestures.

Rösel, R. Die psychologischen Grundlagen der Yogapraxis. Beiträge zur Philosophie und Psychologie. Heft 2. (Stuttgart, 1928).

Roth, Walter Edmund. Ethnological studies among the N.W. Central Queensland aborigines, Brisbane, 1897. xvi, 199 p. Illust. LC
> See especially chap. 4. Elaborate drawings and plates illustrating gestures used by travelling natives forced to use signs because of the variety of spoken languages in Queensland.

——— Additional studies of the arts, crafts, & customs of the Guiana Indians . . . Washington, 1929.

Rudkin, Ethel H. Witches & devils. Folk-Lore, vol. 45 (1934), p. 249-267.
> Gestures used to protect oneself against witches.

Rudolph, H. Der Ausdruck der Gemütsbewegungen des Menschen. Dresden, 1903.

Ruesch, Jurgen and Weldon Kees. Nonverbal communication. Berkeley, California, 1956. Illustrated. Bibliog. p. 197-201. Fla.

> Gestures and other forms of nonverbal communication studied and interpreted from a psychiatrist's point of view. Ruesch was research psychiatrist at the Langley Porter Clinic. He and Kees plead for more research in "nonverbial expression." Their study is stimulating but they fell into the paradox of worshipping at the shrine of non-selective naturalism. Like most pioneering neologists, they sometimes obscure their meaning for the reader, but they may not be overlooked by any student of gestures.

——— Approaches and leave-taking. 16mm film. Running-time about 12 minutes. Langley Porter Clinic, San Francisco, 1955.

Ruesch, Jurgen and Gregory Bateson. Communication, the social matrix of society. New York, 1951. vi, 314 p.

Ruesch, Jurgen. Synopsis of the theory of human communication. Psychiatry, vol. 16 (1953), p. 215-243.

Ruska, Julius. Zur ältesten arabischen Algebra. Der Islam, 1917; and Arabische Texte über das Fingerrechnen, Der Islam, vol. 10 (1920), p. 87-119.

> "M. Ruska résume l'état actuel des études et des textes publiés sur la dactylonomie."—J. G. Lemoine.

Russell, Phillips. Franklin in Paris, The Reader's Digest, 20th Anniversary Anthology. Pleasantville, N. Y., p. 27: Benjamin Franklin greeting French ladies with kiss on the neck.

Russell, William. Rudiments of gesture, comprising illustrations of common faults in attitude and action. Boston, 1838. 120 p. Illust. LC

> Gestures in elocution.

Rutebeuf. Le Miracle de Théophile (13th century).

> See lines 239-242 and 424-425 for gesture of homage of man to devil.

Sachs, C. World history of the dance. New York, 1937. 469 p.

Sacrificial gesture. 'Let the men that sacrifice kiss the calf.'—Hosea, XIII, 2.

Salamalec: see Larousse du XXe Siècle, Paris, 1933, vol. VI, p. 151. Three illustrations of Moslem salaam.

Saluting flag obligatory. Public Law #623, Section 7, passed by the United States Congress June 22, 1942. Later declared unconstitutional by the U. S. Supreme Court, but not before a number of nonsaluters had got into difficulties. See The Reporter (a Pacifist journal), vol. I (June 15, 1943), case of a Jehovah's Witness; also V. W. Rotnem and F. G. Folsom, Recent restrictions upon religious liberty, Bulletin of the American assoc. of university professors, vol. XXIX (April 1943). LC

Salute (military). "When you salute an officer of the U. S. Army, in a very real sense you're saluting God . . ." Chaplain Edward Trower, Hospital Church Bulletin, as quoted in the Fellowship of Reconciliation magazine (July, 1943).

Sanctis, Sante de. Die Mimik des Denkens. Translated by J. Bresler. Halle an der Saale, 1906.

Sand, M. The history of the harlequinade. London, 1928, 2 vols.

Sanlecque, Le P. Louis de. Poème sur le Geste, in Boileau Despréaux, Oeuvres, Troyes, Sainton, 1813, p. 348.

———— Poème sur les Mauvais Gestes, in Abbé Joseph-Antoine-Toussaint Dinouart, L'Eloquence du corps, ou l'Action du prédicateur, Paris, 1761. xx, 447 p. (Contains also Joanne Luca, Actio Oratoris Seu de Gestu et voce libri duo.)

Sartre, Jean-Paul. The age of reason. New York, 1948.
> According to Sartre, drug addicts, prostitutes, and sexual perverts are said to use certain gestures as identification and for solicitation.

Satow, E. Ancient Japanese rituals. Transactions of the Asiatic society of Japan. vol. VII (1881), parts 2 & 4, and IX (1879 & 1881).

Satterthwaite, Linton. (Museum of the University of Pennsylvania).
> On Maya gestures: "modern gestures (of the Maya Indians) have been applied to interpreting the ancient Maya texts—but not very extensively. More should be done . . ." (Personal correspondence)

Sawmill sign language. Lexicon of trade jargon, manuscript in the Library of Congress, Sec. 2, p. 51-54.

Say it with hands; pasimology, or the language of gesture. New York Times magazine (Dec. 22, 1946), p. 49.
> A brief miscellany (two columns) on gestures. Suggest that pasimology, the language of gesture, become an international "bridge" between people.

Schaber, M. De ritibus, vocibus et symbolis salutandi apud populos politos ac humanos antiquorum temporum as nostrae aetatis libelli parten tertiam, qua continetur re ritibus salutandi apud veteres Romanos commentatio. Rastadii, 1858.

Schänzler, Josef. Der mimische Ausdruk des Denkens. Berlin, 1930.

Schechter, Eva I. (Mrs. Abraham I.) Symbols & ceremonies of the Jewish home. New York, 1930, 32 p. LC
> On prayer gestures, see p. 24, 26, and 27.

Schilder, Paul. The image & appearance of the human body. New York, International Universities Press, 1950.

Schinz, Albert. Eighteenth century readings. New York, 1923.
> Louis XIV never passed any woman without removing his hat, chambermaids included. How far he removed hat depended upon the lady's rank. p. 53.

Saluch, Margaret. Recent Soviet studies in linguistics. Science and society. (New York), vol. I (1936), p. 152-167.

Schlosberg, H. Description of facial expressions in terms of two dimensions. Journal of experimental psychology, vol. 44 (Oct. 1952), p. 229-237. Illust. Tables. Bibliog. p. 237.
> Summary of past studies plus suggestions of a new approach.

Schmeller, J. A. Bayerisches Wörterbuch. Munich, 1872-77. 2 vols.
> On gestures used in connection with cunnus, see vol. 2, p. 449 ff.

Schmidt, Leopold. Wiener Redensarten III. Schabab und Schleckabartl. Das deutsche Volkslied, Bd.43, Vienna, 1941, p. 119 ff.

——Die volkstümlichen Grundlagen der Gebärdensprache. Beiträge zus sprachlichen Volksüberlieferungen. Berlin, 1953, p. 233-249.

Schmitz, G. Die Gebärdensprache der Kluniacenser und Hirsauer, Blätter für Taubstummenbildung. Vol. 36 (1923), p. 347-355; 362-364.

Schröder, Franz R. Germanentum und Hellenismus. Heidelberg, 1924. (German. Bibliothek, II, 17)

Schroeder, Karl. Handbuch des Dirigierens und Taktierens. Berlin-Schöneberg, 1937. 10th ed. LC

Schuhl, Pierre-Maxime. Remarque sur le regard. Journal de psychologie normale et pathologique, vol. XLI (1948), p. 184-193.
>Brief philosophical speculation on the variety of facial gestures of men.

Schuler, Edgar A. V for victory: a study in symbolic social control. Journal of social psychology, XIX (May 1944), p. 283-299.
>Author considered the 'V for victory' campaign "the most outstanding in contemporary psychological warfare", discusses its implications, and denounces the advertisers in the USA who "prostituted it for their own gain."

Schultze, V. Zur Geschichte des Händefaltens. Theologisches Literaturblatt, 1892, p. 591.

Schwidetzky, Georg. Do you speak Chimpanzee? (Transl. from the German). London, 1932.

Scorn. See Plutarch, Lives, ed. Harvard Classics, New York, 1910, C. Lentulus, when tried for misusing public funds, held up the calf of his leg, as did boys at ball, when they had missed, and said, "Take this." He was afterward surnamed Sura, "calf-on-the-leg."

Scott, Harry Fletcher and W. L. Carr. Development of Language. New York, 1921.
>See Use of gestures, p. 14-20.

Scott, Hugh L. The sign language of the plains Indians. Archives International folk-lore Association, vol. I (1893), p. 1-206.
>See Seton, Ernest Thompson.

Scott, Robert S. The thumb of knowledge. New York, 1930, XX, 296 p.
>On the Finns' acquisition of magical knowledge by sucking thumb.

Scratching face (for shame, grief). See Libro de Apolonio, ed. C. C. Marden, line 283 C. See also En el mar, Blasco Ibañez: "Pero su mujer no le oìa. Estaba en el suelo, agitada por una crisis nerviosa, y se revolcaba, arañandose el rostro." (The mother has just learned that her son was drowned).

Sealing a bargain: "Echemos, Panza amigo, pelillos a la mar en esto de nuestras pendencias . . ." Don Quijote, chap. 30. 'Echar pelilllos a la mar' is a child's way of sealing a bargain: a hair is pulled from the head, blown upon, as one says, 'Pelillos a la mar.'— Diccionario de la Real Academia Española.

Seemann, Otto. Die Gottesdienstlichen Gebräuche der Griechen und Römer. Leipzig, 1888. 200 p. Illust. LC

> See part I, chap. 5, Das Gebet; illustration of funeral procession gestures, p. 113; part II, chap. 4, Gebet und Fluch; et passim.

Seligmann, C. G. & A. Wilkin. The gesture language of the Western Islanders. Cambridge Anthrop. Exp'd. to Torres Straits Reports, vol. 3 (1907), p. 255-260. NYP

Seligmann, Siegfried. Der böse Blick und Verwandtes. Berlin, 1910. 2 vols. Illust. LC

> On phallic signs, vol. II, p. 192; 202; on cunnus, II, 204; on the rump, I, 174.

—— Die Zauberkraft des Auges und das Berufen. Hamburg, 1922. xxxix, 556 p. Illust. LC

> An exhaustive and heavily documented study of the evil-eye, and of its opposite the GUTE BLICK, as they were known down the centuries around the world. A fundamental work.

—— Die magischen Heil- und Schutzmittel aus der unbelebten Natur, mit besonderer Berücksichtigung der Mittel gegen den Bösen Blick; eine Geschichte des Amulettwesens. Stuttgart, 1927. xi, 309 p. 111 illustrations. LC

> Bibliography at end of each chapter. See index under Böser Blick, Guter Blick, Gesten (magische), Mano cornuta, mano fica, etc.

Seneca. Moral essays. London, 1928, 3 vols. John W. Basore, translator.

> See vol. I, book 3, De ira, for remarkable description of the gestures of an angry man.

Seton, Ernest Thompson. Sign talk; an universal signal code without apparatus, for use in army, navy, camping, hunting and daily life. The gesture language of the Cheyenne Indians with additional Signs used by other tribes, also a few necessary Signs from the code of the Deaf in Europe and America, and others that are established among our Policemen, Firemen, Railroad Men, and School Children, in all 1725, prepared with assistance for General Hugh L. Scott. 700 Illust. by the author. Garden City, New York, 1918. LC

> The French and German equivalent words are added after the English.

Sex gestures and postures.

> Currently stock-in-trade in American advertisements. Also many "girlie" magazines specialize in erotic gestures and postures.

Shaftesbury, Edmund. Lessons in acting. ca. 1885.
> "A . . . startling example illustrating codified pantomine . . . ; it not only lists and describes an exhaustive catalog of bodily attitudes alleged to symbolize every conceivable mood and emotion, but provides no less than 100 line cuts to illustrate them."—John Dolman, Jr., The art of acting, q. v.

Shannon, B. What do your hands reveal. World today, vol. 53 (April, 1929), p. 462-466.

Shawn, Ted. Every little movement; a book about François Delsarte, the man and his philosophy, his science and applied aesthetics, the application of this science to the art of the dance, the influence of Delsarte on American dance. Pittsfield, Mass., 1954. 115 p. Charts. Fla.

Shea, George E. Acting in opera, its A-B-C, with descriptive examples, practical hints and numerous illustrations. London, 1915. LC

Siddons, Henry. Practical illustrations on gesture and action. 1807.
> First ed. with 66 full-page engravings, many showing well-known actors in famous parts.

Si-Do-In-Dzou. Gestes de l'officiant dans les cérémonies mystiques des sectes Tendai et Singon d'après le commentaire de M. Horiou Toki. Traduit du Japonais par S. Kawamoura, avec introduction par L. de Milloué. Musée Guimet (Annales), Bibliothèque d'étude, VIII, Paris, 1889. NYP
> Thoroughly documented, illustrated with several hundred drawings. Explanations of mystical, and other, meanings given. Introduction by L. de Milloué, p. i-xix. Illustrated gestures primarily manual.

Sign of cross: see P. Graff; Hastings; Catholic Encyclopedia; Philippe de Thaün, Bestiare, 'the Ibis' (12th century); J.-R. Chevaillier and Pierre Audiat, Les Textes Français, Paris, 1927, p. 153, note on chap. 25 of Rabelais' Gargantua (sign of cross with left hand tabu); Montaigne, Essays, I, chap. 56; Juan Ruiz de Alarcón, Anticristo, Act I, final scene.

Signals . . . the secret language of baseball in finger-tip movies. Boston, 1957. Illustrated.
> The material in this booklet is based on a series of articles that appeared in a national baseball weekly.

Sittl, Karl. Die Gebärden der Griechen und Römer. Leipzig, 1890. 386 p. Illustrated. LC

> Indispensable. All gestures described are documented and there are thousands of them. See, for example, Monastic signs, p. 224; phallic, p. 100-101, 121-125 (fig. 7); cunnus, p. 123; the rump, p. 124; etc.

Skraup, K. Katechismus der Mimik und Gebärdensprache. Leipzig, 1892.

Slapping one's forehead (to express amazement). Cervantes, Don Quijote, ed. F. Rodrìquez Marín, Madrid, 1952, vol I, p. 13. Same gesture found in Jose Marmol, Amalia, Montevideo, 1851, vol I, p. 174-175. The gesture is very common today in the Hispanic world.

Smith, D. E. and Karpinski, L. C. The Indu-arabic numerals. Boston, 1911.

> "C'est l'ouvrage qui fait actuellement autorité; il présente le dernier état de la question sur l'origine des chiffres."—J. G. Lemoine.

Smith, Logan Pearsall. Words and idioms. Studies in the English language. 2nd edition. London, 1925. See especially p. 249 ff. and Appendix, p. 279-292. Fla.

Smith, Stella Tilley. Imagery of motion in Shakespeare's tragedies. Ph.D. thesis, University of Florida, 1955. Unpublished.

Snarl. "Come, come, girls, take it easy." Associated Press wirephoto of two young women fighting over a man; remarkable picture of snarls of violent anger in human beings. Greensboro (North Carolina) Daily News (July 23, 1957).

Sorrel, Walter. Your gestures give you away. American magazine, vol. 140 (Nov., 1945), p. 140. Illustrated.

> The author is a consulting psychologist. He makes some unqualified claims about the secret meaning of nine common nervous gestures.

Spaier, A. La pensée concrète. Paris, 1927. 466 p. Bibliography p. 431-440. Fordham University library.

Spalding, A. C. B. Portal to rhetorical delivery, with questions, exercises and observations on the New System of corporal expression. Dublin, 1826. 224 p.

> Elocution and gesture.

Speck, Frank G. and others. Cherokee dance & drama. Berkeley, California, 1951. XV, 106 p. Illustrated.
>See passim; for example, the 'corn-dipping gesture', p. 77.

Speech: platform posture and movement. 16 mm film. Young America, 1949. 11 minutes. Centron Corporation.
>On posture and movement of public speakers.

Speech: function of gestures. 16 mm. Young America, 1950. 11 minutes. Centron Corporation.
>The function of gesture in public speaking.

Spitting (to express scorn): Job XVII,6; Seneca, De ira, book III, chap. 38 (extreme example); Shakespeare, Richard the Third, act I, sc. 2, lines 145-147.

Spreen, Hildegarde L. Folk-dances of South India. Madras, 1948. xvi, 134 p.
>Illustrated and documented. Shows a wealth of gracious and meaningful gestures.

Sprinkling dust on head (grief): Job II, 12-13.

Stählin, W. Vom Sinn des Leibes. Stuttgart, 1934.

——— Form und Gebärde im Gottesdienst und Gebet. Frauenhilfe, 1940. (Off-print)

Staley, Vernon (editor). The Library of liturgiology and eccesiology: essays on ceremonial. London, 1904, Vol. 4, 307 p. Illustrated.
>See chapter I, English ceremonial; the blessing, p. 29, 70, 71; sign of cross, 29, 255 et passim; beating the breast, 255. See index for other ritualistic gestures.

Stamping foot: Plutarch, Lives, Harvard Classics, New York, 1910, p. 303-304: chapter on Caesar.

Stand (praying): Mark XI, 25; 1 Chronicles XXIII, 30.

Stanistreet, G. M. Pantomine is easy. Recreation, vol. 38 (May-June 1944), p. 72-74, 137-138. Illustrated.
>Pantomimes for children . . . stories enacted . . . 'Sleeping Beauty', etc.

Stearns, Jo Ann. How to speak with the hands. Mexico this month, vol. II (July 1956), p. 22-23. 4 illustrations.
>A brief description of the personal experience of an American girl with gesturing Mexicans.

Stebbins, Genevieve. Delsarte system of expression 6th edition. New York, ca. 1901. 507 p.
> Illustrated with line drawings, photos of Greek art, etc. Alarmingly complete treatise on gestures, highly 'aesthetic', often obscure. Fla.

Steinhausen, G. Der Gruss und seine Geschichte. Kulturstudien, vol. (1893), p. 1-17.

Steinitzer, Max. Die menschlichen und tierischen Gemütsbewegungen als Gegenstand der Wissenschaft. Munich, 1889. (Giebt eine rein historische Darstellung der Theorien des 15-18 Jahrhunderts)

Stevens, Henry Harmon. Description in the Hellenic dramas of Franz Grillparzer. Ph.D. Diss., Harvard University, 1916.

Stirling, E. C. Report on the work of the Horn scientific expedition to Central Australia. London and Melbourne, 1896, Part IV, p. 111-125. Illust.
> The Arunta sign language.

Stoebe, K. Altgermanische Grussformen. Paul und Praunes Beiträge zur Geschichte der deutschen Sprache und Literatur. Vol. XXXVII, p. 173 ff.

Storfer, A. J. Marias jungfräuliche Mutterschaft. Berlin, 1914. (Neue Studien zur Geschichte des menschlichen Geschlechtslebens, I)
> On phallic signs, p. 35.

Strange radio sign talk directs radio broadcasts. Popular mechanics, vol. 58 (July 1932), p. 25. Illust.
> Very brief.

Straus, Erwin W. Der Seufzer; Einführung in eine Lehre vom Ausdruck. Jahrbuch für Psychologie und Psychotherapie, II (1954), p. 113-128.
> The sigh serves no clearly physiological or psychological function, but it is certainly an expressive movement.

Strehle, H. Analyse des Gebärdens, Körpergewegung und-Haltung. Industrielle Psychotechnik, XI (1934), p. 89-90. LC

Stroke beard (to show affection)
> Mullah Kashani, Iranian spiritual leader, stroked beard of youthful murderer of Premier Ali Razma to show his affection and approbation. Photo in Life Magazine (Dec. 8, 1952). p. 52.

Strowski, F. Théâtre et nous Paris, 1934. 219 p. LC
>Chap. IV offers brief philosophical discourse on all gestures in general and actor's gestures in particular, with special reference to the silent movies.

Sullivan, John F. The externals of the Catholic Church: her government, ceremonies, festivals, sacramentals, and devotions. New York, 1918. xi, 385 p. Illust. Fla.

Sulzberger, C. L. Ethiopia approves area defense pact, New York Times, (Dec. 21, 1952).
>Offers specific instructions on gestures and postures necessary when entering or leaving the presence of Haile Selaisse, the Elect of God.

Summers, Montague. The history of witchcraft and demonology. New York, 1956. 353 p. 2nd ed. Fla.

Surgeon's gestures.
>To the instrument-nurse the surgeon signals for the instrument needed with a hand code.

Surrealistic gesture. See Coronet magazine (Jan. 1940), p. 33.

Szymanski, J. S. Aktivität und Ruhe bei den Menschen. Zeitschrift für angewandte Psychologie, vol. XX (1920), p. 192-222.

Taboureau, Jean. Je lis dans les gestes; démarches, tics, mimiques. Paris, 1938. 80 p. Illust. LC

Taladoire, Barthelemy A. Commentaires sur la mimique et l'expression corporelle du comédien romain. Montpellier, 1951. 138 p. Thèse complémentaire, Paris; bibliog. p. 133-136. Col. Univ. Lib.
>Well-documented study of Roman and Greek gestures, particularly those of actors and speakers.

Tap palm of hand of another (seal bargain)
>Sheep merchant, Dingdong, taps palm of Panurge's hand in Rabelais, Pantagruel, Book IV, chap. 16; merchant remarks: "Fourchez la"

Tatlock, John S. P. and A. G. Kennedy. A concordance to the complete works of Geoffrey Chaucer and to the Romaunt of the Rose. Washington, 1927. Fla.
>See under embrace, hand, etc.

Taylor, Archer. Folklore and the student of literature. Pacific spectator, vol. II (1948), p. 216-223. Fla.
>On folk gestures see esp. 217-218.

——— The Shanghai gesture. FF Communications No. 166, Helsinki, 1956. 76 p. Bibliography. Fla.

> A thorough study of the gesture of thumbing-the-nose, its age, its spread, its names in a number of languages, especially English.

——— The Judas curse. American journal philology vol. XLII (1921), p. 234-252. Fla.

> See p. 251.

Taylor, Mark R. Norfolk folklore. Folk-lore (London), vol. 40 (1929), p. 113-133.

> On the evil eye, see p. 126, 129.

Tchang Tcheng-Ming. L'écriture chinoise et le geste humain. Essai sur la formation de l'écriture chinoise. Paris, 1937. 205 p. Reproduction of 700 ancient characters.

> One of the fundamental works of gesture study.

Teeth (gnashing) and hissing: Lamentations, II, 16.

Teit, James A. The Salishan tribes of the Western Plains; the Coeur d'Alêne. 45th Ethnological report, Wash., D. C., 1928. U. S. Government Printing Office.

> One section deals with Indian sign-language.

Television-studio sign language.

> See almost any television manual; for example, Art Gilmore and G. Y. Middleton, Television and radio announcing, Hollywood, 1949, chap. I. Illust.

Telford, Ira R. An anatomist looks at your hand. The courier of George Washington University Medical Center, VII (Mar. 1955), p. 10-13. 3 illustrations.

> A professor of anatomy romanticizes on the hand.

Thomas, Adolph V. L'Anthropologie du geste et les proverbes de la terre. Revue Anthropologique (Oct.-Dec. 1941), p. 164-94. Bibliog. p. 193-194. LC

> Supports theory of gestural origin of language. Offers complementary evidence to that of M. Jousse, q. v. Cites Jousse's evidence gathered from observation of contemporary 'primitive' farmers in France and peoples from other parts of the world. Draws a parallel between gesture and proverb.

Thomas, Lawrence L. The linguistic theories of N. Ja. Marr. University of California publications in linguistics, vol. XIV (1957), p. 1-176. Bibliography.

> See especially p. 97 ff. Maintains that manual language of gesture preceded oral language by many thousands of years. Conclusions, p. 135-146.

Thomas, P. Hindu religion, customs, and manners. D. B. Taraporevala sons & Co., Ltd., Bombay, n.d. 2nd ed. Illust.
> Religious gestures of gods and worshippers predominate. See p. 80 for greeting gestures.

Thomasius, Jacobus. De Ritu Veterum Christianorum Precandi versus Orientum. Dissertation, Leipzig, 1670.
> Cited by Franz J. Dőlger in "Sol Salutis . . . ", p. vii.

Thompson, John Eric S. Maya hieroglyphic writing; introduction. Washington, 1950. xvi, 347 p. Illust. Fla.
> In illustrations of Maya writing, note widespread use of gestures by ancient Mayas.

——— Rise and fall of Maya civilization. Norman, Oklahoma, 1954. xii, 287 p. Illust. Fla.

Thompson, Stith. Motif-index of folk literature. Bloomington, Indiana, 1932-37. (A new edition is in progress)
> See under hand, finger, etc.

Thomson, J. A. Human hand. New statesman, vol. 16 (Jan. 15, 1921), p. 443-4.
> Speculation on the hand.

Thorek, Max. The face in health and disease. Philadelphia, 1946. 781 p. 636 illustrations. Bibliog. LC
> Author's point of view is that of a surgeon, but the work contains numerous items and observations of interest to students of gesture. See, for example, chap. 5: 'The face in emotional and physical states.' Author hopes to stimulate more study of the face as a diagnostic guide which may register the condition of the body and the mind.

Thousand and one nights: see The book of the thousand nights and one.

Thumbing a ride. 'It's the patriotic gesture now!' Life magazine (July 6, 1942).
> Large photo of four people of varying ages and social status thumbing a ride together. Full-page advertisement.

Thumbs up. Widely used during World War II. For an example, see photo in Greensboro (North Carolina) Daily News (Dec. 9, 1941). Caption: "Those upheld thumbs mean, 'Let me at those Japs.'" Photo shows group of enlistees, each one making the thumbs-up signal.

Thumbs up (and down). Cartoon wholly of drawings of thumbs up and down, Florida Times Union (Feb. 28, 1957) Caption: "Big three unity—all thumbs."

Tikkanen, Johan Jakob. Zwei Gebärden mit dem Zeigefinger. Druckerei der Finnischen Literaturgesellschaft. 1913. 107 pp. Finska Vetenskaps—Societeten. Acta Societatis Scientiarum Fennicae. Tomus 43, #2.). Illust. Bibliog. p. 102-107. NYP

> Thoroughly documented from art of many centuries. Primarily concerned with two hand-gestures in art.

Tipping-hat tabu. "It is a social error to tip one's hat to a Moslem girl."—Dr. M. Yusuf Khan, as reported in the Greensboro (North Carolina) Daily News (June 6, 1943).

Tissié, Philippe. La science du geste. Revue scientifique, ser. 4, vol. 16 (1901), p. 289-300. LC

> Curious speculation on the 'beauty' of gestures by a physical educationist who declaims that "L'education physique n'est pas un moyen athlétique: c'est une fin philosophique."

Tobacco auctioneer gestures. Greensboro (North Carolina) Daily News (Sept. 29, 1944). See also painting of similar gestures in full-page advertisements of Lucky Strike cigarettes in national magazines in U.S.A. in 1942.

Tomkins, William. Universal Indian sign language of the Plains Indians of North America, together with a simplified method of study, a list of words in most general use, a codification of pictographic symbols of the Sioux and Ojibway, a dictionary of synonyms, a history of sign language, chapters on smoke signalizing, use of idiom, etc., and other important co-related matter. San Diego, California, 1954. 11th edition. 106 p. (1st ed., 1926) LC

> French and Indian equivalents are given with each illustration. The author lived as a boy (1884-94) on the edge of the Sioux Indian reservation in Dakota Territory.

Touch Testicles (insult).

> When the crowd hooted at a bullfighter in Mexico because of unsatisfactory performance, he insulted them by touching his trousers at the crotch. The event was indignantly written up in the local press; the following day he apologized in the bull-pen with the appropriate gestures directed to the crowd.

Train-crew gestures. 'Cross country hot shots—story of freight-train's crew.' Look magazine (Dec. 23, 1941), p. 37-41.

Tremearne, A. J. N. Hausa superstitions and customs. London, 1913. 548 p. Illust. LC
> See index under evil eye, gesture, etc.

Twain, Mark. In sketches new and old; see The petrified man for burlesque use of double Shanghai gesture.

Twining, Louisa. Symbols and emblems of early and mediaeval Christian art. London, 1852. xii, 190 p. 93 plates. LC
> See plate 1, no. 14; plate II, 16, 19.

Tykulsker, P. Reference to the face in French drama before Racine. Modern language notes, vol. 51 (June 1936), p. 381-386.
> A variety of allusions to facial gesture found in French literature before Racine.

Tylor, E. B. Primative culture. London, 1873. 2 vols. Fla.
> See chap. on 'The art of counting.'

——— Researches into the early history of mankind. New York, 1878 (3rd edition). 388 p. Fla.
> Contains information on greetings, such as nose-rubbing, blowing on; prayer gestures; gestures of the Cistercian monks, etc.

Tyra de Kleen: See Kleen

Udine, J. d' (pseud. Albert Cozanet). L'art et le geste. Paris, 1910.

Ullman, B. L. Ancient writing and its influence. New York, 1932.
> Says that theory of Egyptian origin of Phoenician alphabet has been discarded by many. Origin of alphabet unknown. "In picture writing there is much imitation of gesture — a form of communicating ideas very common among primitive peoples." — p. 6-7. In personal correspondence Ullman writes: "I think that the ancestor of our letter E, Hebrew *he*, could be regarded as a gesture sign. The word *he* means 'behold' in Hebrew and this led Sethe to suggest as the origin of this sign the Egyptian figure of a man with upraised hands. This suggestion of Sethe, made perhaps a generation ago, has been generally accepted."

Urges sign language to promote world peace. Hobbies (Aug. 1952), p. 137.
> Mostly quotations from Wm. Tomkins, q. v., who urges Indian gestures as a common world-wide medium of communication. Sign language served the North American Indians for thousands of years; so why not promote its universal usage among the nations today? At world jamboree of boy scouts in England in 1929, it was a common sight to see a Zulu scout talking to a German or Russian or Chinese scout ... using Indian sign language.

Urtel, Hermann. Beiträge zur portugiesischen Volkskunde. Abhandlungen aus dem Gebiet der Auslandskunde. Band 27. Hamburg, 1928. VIII, 82 p. 4 plates. Bibliographical footnotes.

> Contains chap. on "Zur Gebärdensprache", p. 4-22; also study of the figa in the chap. on "Amulette", p. 23-26. See esp. Tafel 2.

V for Victory. See E. A. Schuler in this bibliography; also Current history, June 1941, p. 28, on the British campaign of V for Victory.

Valentini, F. Plates from Trattato Su La Commedia dell' Arte ossia Improvisa Maschere Italiane ed Alcune Scene de Carnevale de Roma. Berlin, 1826. First ed.

> A set of 20 colored plates showing the major characters of the commedia dell' arte, including Arlecchino, Pantalone, Scopetta, etc.

Vaschide, N. Essai sur la Psychologie de la Main. Paris, 1909.

> Takes up the hand in chiromancy, in art, finger prints, from the anthropological viewpoint, pathology of the hand, etc. Highly speculative.

Vavrouy: See Boyvin de Vavrouy.

Vega, Lope de. El remedio en la desdicha, II, final lines: handshake to seal bargain; El acero de Madrid, II, page 185 (in ed. of Biblioteca de Autores Españoles), servant makes sign of fig at another man with hand of girl; El cabellero de Olmedo, III, sc. 4: on laughter; El divino africano, Act I: stage directions indicate that students in the class portrayed shuffle and stamp their feet (patean) to indicate they are anxious to leave because the day is a holiday.

Vendryes, Joseph. Language. A linguistic introduction to history. New York, 1925. 378 p. Translated from the French. Bibliography. Fla.

> See esp. part 5 of chap. I on origin of writing.

Venkatachalam, G. Dance in India. Bombay, Nalanda Publications, 194?. 131 p. Illust. LC

> The many illustrations speak for themselves to emphasize the central place gesture holds in the dance in India. The accompanying text is that of an enthusiast. Numerous hand gestures are illustrated and explained, p. 125-31. Exemplary quotation: "The gesture language is as old as man; but only (sic) in India it has been developed and perfected (sic) to play as important a role as the spoken language not only in rituals and temple worship but in the arts of drama and dance." Mudras, occuring frequently in the text, means 'hand gesture.'

———. Srimati Shanta, Bharata Natyam. Bangalore, 1944. iv, 102 p. Illust. In English. LC

> Random comments on gesture and dancing in India.

Verplanck, W. S. Panel on psychology and physiology. A survey report on human factors in undersea warfare. Washington, National Research Council, 1949.

> See section on gesture languages.

Verrill, Ruth. (Chiefland, Florida): re Maya gestures:

> In personal correspondence Mrs. Verrill, an artist and scholar acquainted with the Maya Indians, expresses the belief that search of oriental gestures would clarify the meaning of some of the numerous Maya gestures found portrayed on remains, especially ritual, ceremonial, and symbolical hand-signs. See A. H. Verrill and Ruth Verrill, America's ancient civilizations, New York, 1953, chap. 4. With respect to linguistic relationships between aboriginal Indian languages and the Orient, see F. Pérez de Vega, Las lenguas aborígenes, Caracas, Editorial Daily Journal, 1957, 112 p.

Victory gesture (Biblical): I Samuel XV, 12; 2 Samuel XVIII, 18.

Video director; hand signals used in studios. New York Times Magazine (Oct. 24, 1948), p. 58.

> One column, illustrated, on meaning of some gestures used in television studios.

Vierordt. Das Händefalten im Gebet. Theologische Studien und Kritiken, vol. 26 (1853), p. 89-93. LC

——— De junctarum in precando manum origine Indo-Germanico et usu inter plurimos Christianos adscito. Cum tabula litographica. Karlsruhe, 1851.

Vignes-Rouges, J. des. Les révélations du visage. Les annales politiques et littéraires. Vol. 109 (Feb. 1937), p. 183-185.

Villiers, Elisabeth and A. M. Pachinger. Amulette und Talismane. München, 1927.

> Extensive collection of amulets with some indicating hand gestures.

Virgil: See Monroe N. Wetmore, Index verborum.

Visscher, II. Religion und sociales Leben bei den Naturvölkern. Bonn, 1911. Duke University Library.

> See vol. I, p. 113 ff, on primitive gestures.

Vlašimsky, J. Mimische Studien zu Theodor Storm. Euphorion, vol. XVII, p. 636; XVIII, 150, 468 ff.

von Wied, Prinz. Reise in das Innere von Nordamerika, 1832-34. Koblenz, 1841, vol. II, p. 645-653.

Voorhees, Oscar M. The history of Phi Beta Kappa. New York, 1945.
> See the sign of the society, discussed p. 10; drawing illustrating secret handshake, p. 135.

Vorwahl, H. Die Gebärdensprache im Alten Testament. Berlin, 1932.

—— Die Gebärdensprache der Religion. Zeitschrift für Religionspsychologie 5, 1932, p. 121-128.

Vossler, K. Geeist und Kultur in der Sprache. Heidelberg, 1925.

Voullième, E. Quomodo veteres adoraverint. Dissertation, Halle, 1887.

Vuillier, Gaston. A history of dancing, from the earliest ages to our own times. London, 1898. 446 p. Profusely illustrated. LC
> Ancient and modern sources for illustrations. Numerous meaningful gestures shown to have been used in the art of miming and in the dance.

Wagner, M. L. Phallus, Horn und Fisch. Jabergfestschrift Romanica Helvetica, IV, Zürich und Leipzig, 1937.

Walker, Jerrell R. Sign language of the Plains Indians. Chronicles of Oklahoma, vol. 31 (1953), no. 2, p. 168-177. NYP

Walpole, Hon. F. T. The Ansayrii and the Assassins. London, 1851.
> Contains one short paragraph on the Ansayrii signs and salutes of recognition (vol. III, p. 354)

Wandruska, Mario. Haltung und Gebärde der Romanen. Beihefte zur Zeitschrift für romanische Philologie. (Tübingen), vol. 96 (1954), p. 1-100. Bibliographical footnotes.
> Well-documented account of la contenance, el contegno, le port, l'atto, la posture, le geste, l'allure, l'air, la manière, la gravité and la grâce.

Ward, J. S. M. The sign language of the mysteries. London, 1928. 2 vols. Illust. NYP
> Author is a Freemason. Holds sign language of great significance in the world during many centuries until "the Renaissance and Reformation to a large extent shattered the old symbolic system" (p. iii). Author maintains that "far back in the dawn of time there was a basic Sign Language common to all, or nearly all, the members of the human race," and that this language has survived to a far greater extent than has been suspected. He supports this theory by indicating a large number of gestures around the world, and down the centuries, which are of common meaning (p. 3). See especially chronological listing of signs, vol. II, p. 193-237; Index, p. 242-245.

———— The Hung society. London, 1925. 3 vols. Illust. LC
> Complex system of gestures and postures of the Hung society. See esp. vol. I, chap. 10. Bibliography and index.

Warman, Edward B. Gestures and attitudes; an exposition of the Delsarte philosophy of expression, practical and theroretical. Boston, 1892. 422 p. 154 illustrations. LC
> For elocutionists and speech teachers.

Waving handkerchiefs en masse (Russia).
> At the close of services in a Baptist church in Moscow visited by a group of Quakers, the congregation sang "God be with us til we meet again" in Russian and while singing all the women took out their handkerchiefs and waved them to the visitors. — Kathleen Lonsdale, Quakers visit Russia, London, ca. 1951, p. 26.

Wenger, M. A. An attempt to appraise individual differences in level of muscular tension. Journal of experimental psychology, vol. XXXII (1943), p. 213-225.

Weiss, Paul. The social character of gestures. The philosophical review, vol. LII (1943), p. 182-186.
> A speculative debate (about gestures) with George Mead.

Wespi, Hans Ulrich. Die Geste als Ausdrucksform und ihre Beziehungen zur Rede, Darstellung anhand von Beispielen aus der französichen Literatur zwischen 1900 und 1945. Zurich, 1949. xx, 58 p. Thesis. Bibliog. p. xiii-xx. LC
> 'Die vollständige Arbeit erscheint als Band 33 der Romanica Helvetica, series linguistica.'

Westermarck, Edward. Ritual & belief in Morocco. London, 1926. Illust.
> See vol. I, xxxii, 608; vol. II, xvii, 629 ff. Also index under evil eye, and passim.

Wetmore, Monroe N. Index verborum Vergilianus. New Haven, 1911.

Whitmire, Laura G. A course in pantomime. Quarterly journal of speech education, vol. XIII (1927), p. 110-118.
> Advice on teaching pantomime, suggested skits, bibliog.

Wiegand, J. Die Gesten in der deutschen erzählenden Dichtung. Neue Jahrbücher für das klassische Alterlum . . . , vol. 40 (1918), p. 332-44. Duke Univ. Library

Wilfing, Jutta. Etwas über klassische Gebärdensprache und was davon mit unseren Redensarten zusammenhängt. Aus alten Aufzeichnungen gesammelt. Der getreue Eckart, XIII (1935-36), p. 122 ff.

Wilhelm. Klagebärdenbilder und Klagegebärden in der deutschen Dichtung des höfischen Mittelaters. Diss. Bonn, 1936.

Wiseman, Cardinal. Essays on various subjects. London, 1855. Vol. III, p. 533-555.
> On Italian gestures.

Wissowa, Georg: See Pauly, A. F. von.

Witchcraft. For magic gestures used in witchcraft, see almost any standard volume, v. g., Summers; Margaret Murray; Hole, etc. The mass of material on the subject of witchcraft is vast.

Witte, O. Untersuchen über die Gebärdensprache. Beiträge zur Psychologie der Sprache. Zeitschrift für Psychologie. vol. CXVI (1930), p. 225-308.

Wolff, Charlotte. A psychology of gesture. (Translated from the French manuscript by Anne Tennant). London, 1948. xvii, 225 p. Illustrated with 44 plates. Bibliography, p. 214-217, mainly items from psychology. LC
> The author's wish is to indicate a method of intepretation of subconscious gestures. Emphasis is on hand gestures. Some of the chapter headings are 'The physiology of gesture,' 'The pathology of gesture,' 'The natures of gesture,' etc. Often speculative.

———— The human hand. London, 1943. xvii, 198 p. 24 plates. Bibliography p. 193-198. LC
> This book is called 'the psychology of the hand.' It is based on a psychophysiological theory, published in outline in the British Journal of Medical Psychology, and contains a 'new' method of hand-interpretation, resulting from the theory and from statistical research on normal and abnormal subjects carried out by the author at University College, London, and in a number of clinics and hospitals in Paris and England.

———— The hand in psychological diagnosis. London, 1951. 218 pp. Illustrated. Bibliog. p. 211-214. LC
> Speculation on psycho-motor phenomena; accounts of research on significance of types of hands of normal and abnormal persons; relationship between the psyche and the hand. Sample quotation: "It is safe to say that there is practically no mental activity in which the hand is not in one way or another involved."

―――― Expression: excerpts. Review of reviews (London), vol. 85 (Feb. 1934), p. 37.

Wolfram, Richard. Die Volkstänze in Oesterreich und verwandte Tänze in Europa. Salzburg, 1951. LC

Woodman, C. M. Quakers find a way. New York, 1950.
>See p. 228-9 for account of 17th century Quakers who refused to remove hats before the court.

Wundt, Wilhelm. Elements of folk psychology. London, 1916 et seq. (Translated from the German) LC
>Much quoted on gestures of primitive man, primitive gesture 'syntax', etc.

Wuttke, Adolf. Der deutsche Volksaberglaube der Gegenwart. Bearbeitung von E. H. Meyer. Berlin, 1900. xvi, 535 p. 3rd ed. Bibliog. p. xiv-xvi. LC
>See under Der böse Blick, et al, paragraph 220; in index see under Segen, Hand, grüssen, etc.

Yahraes, Herbert. What do you knok about blindness? New York, 1947. 32 p. Illust.
>See p. 19-20: 'Keeping a person from looking blind.' The blind should learn the gestures and postures and movements of sighted persons; otherwise, they appear too still.

Yerkes, Royden K. Sacrifice in Greek and Roman religions and early Judaism. New York, 1952. xix, 267 p.
>Sign of the cross, p. 143; laying on of the hands, p. 133 ff. See also passim for other gestures.

Young, Robert. Analytical concordance to the Bible. New York, 1936.
>See under hand, embrace, greet, kiss, salute, bless, head, etc. A fundamental source.

Your gestures give you away. Science digest (Oct. 1955), p. 30. Illust.

Zappert, G. Ueber den Ausdruck des geistigen Schmerzes im Mittelalter. Denkschriften der philosophisch-historischen Klasse der Akademie der Wissenschaften zu Vienna. V, p. 73. LC

Zeidler, M. Melchior. Exercitatio Theologica de Conversione Orantium, ceu Ritu Ecclesiae iam Olim Usitato. Dissertation. Königsberg, 1673.

Ziff, W. B. Your gestures give you away. Reader's digest (Aug. 1951), p. 104-106. Illust.
> Unreliable. Typical quote: "The overquick handshake is that of an incorrigible self-seeker." On this basis all Frenchmen and most of the people of the Near East, not to mention Latin America, are incorrigible self-seekers.

Zola, Emile. Germinal, part 5, chap. 5: gross gestures of Monquette described.

Zons, F. B. Von der Auffassung der Gebärde in der mittelhochdeutschen Epik. Dissertation. Münster, 1934.

Zschietzschmann, Willy. Untersuchungen zur Gebärdensprache in der älteren griechischen Kunst. Dissertation, Jena, 1924.

Zung, Cecilia S. L. Secrets of the Chinese drama. A complete explanatory guide to actions and symbols as seen in the performance of Chinese drama, with synopses of 50 popular Chinese plays and 240 illustrations, including 54 colour plates. Shanghai, Hong Kong, Singapore, 1937, xv, 299 p. LC
> Part 2, Technique, contains chapters on the movements of each of the following: sleeve, hand, arm, foot, leg, waist; also pheasant feather movements and some symbolic actions. Sleeve movements are the most important. All gestures are explained.

Zunz, Leopold. Der Ritus Synagogalen Gottesdienst. Berlin, 1859.
> Jewish liturgy and ritual.

University of Florida

BODY MOVEMENT
Perspectives in Research
An Arno Press Collection

Anthropological Perspectives of Movement
  a. LaBarre, Weston
    The Cultural Basis of Emotions and Gestures (Reprinted from *Journal of Personality,* Vol. 16, 1947)
  b. Bailey, Flora L.
    Navaho Motor Habits (Reprinted from *American Anthropologist,* Vol. 44, 1942)
  c. Hewes, Gordon W.
    World Distribution of Certain Postural Habits (Reprinted from *American Anthropologist,* Vol. 57, 1955)
  d. Kurath, Gertrude Prokosch
    Panorama of Dance Ethnology (Reprinted from *Current Anthropology,* Vol. 1, May 1960)
  e. Hall, Edward T.
    A System for the Notation of Proxemic Behavior (Reprinted from *American Anthropologist,* Vol. 65, 1963)
  f. Hayes, Francis
    Gestures: A Working Bibliography (Reprinted from *Southern Folklore Quarterly,* Vol. 21, 1957)

Christiansen, Bjørn
Thus Speaks the Body: Attempts Toward a Personology from the Point of View of Respiration and Postures. Oslo, Norway, 1963

Davis, Martha
Towards Understanding the Intrinsic in Body Movement. New York, 1974

Dewey, Evelyn
Behavior Development in Infants: A Survey of the Literature on Prenatal and Postnatal Activity 1920–1934. New York, 1935

Evolution of Facial Expression: Two Accounts
  a. Andrew, R. J.
     The Origin and Evolution of the Calls and Facial Expressions of the Primates (Reprinted from *Behaviour,* Vol. 20, Leiden, Netherlands, 1963)
  b. Huber, Ernst
     Evolution of Facial Musculature and Facial Expression. Baltimore, 1931

Facial Expression in Children: Three Studies
  a. Washburn, Ruth Wendell
     A Study of the Smiling and Laughing of Infants in the First Year of Life (Reprinted from *Genetic Psychology Monographs,* Vol. 6, Nos. 5 & 6, Worcester, Mass., 1929) November–December, 1929
  b. Spitz, René A., with the Assistance of K. M. Wolf
     The Smiling Response: A Contribution to the Ontogenesis of Social Relations (Reprinted from *Genetic Psychology Monographs,* Vol. 34, Provincetown, Mass., 1946) August, 1946
  c. Goodenough, Florence L.
     Expression of the Emotions in a Blind-Deaf Child (Reprinted from *Journal of Abnormal and Social Psychology,* Vol. 27, Lancaster, Pa., 1932)

Psychoanalytic Perspectives of Movement
  a. Ferenczi, S.
     Psycho-analytical Observations on Tic (Reprinted from *The International Journal of Psycho-Analysis,* Vol. 2, March, 1921)
  b. Feldman, Sandor
     The Blessing of the Kohenites (Reprinted from *American Imago,* Vol. 2, 1941)
  c. Kris, Ernst
     Laughter as an Expressive Process: Contributions to the Psycho-Analysis of Expressive Behaviour (Reprinted from *International Journal of Psycho-Analysis,* Vol. 21, 1940)
  d. Mahler, Margaret Schoenberger
     Tics and Impulsions in Children: A Study of Motility (Reprinted from *Psycho-Analytic Quarterly,* Vol. 5, 1944)
  e. Deutsch, Felix
     Analytic Posturology (Reprinted from *Psycho-Analytic Quarterly,* Vol. 21, 1952)
  f. Braatoy, Trygve
     Psychology vs. Anatomy in the Treatment of "Arm Neuroses"

with Physiotherapy (Reprinted from *Journal of Nervous and Mental Disease,* Vol. 115, 1952)
g. Mittelmann, Bela
Psychodynamics of Motility (Reprinted from *Psychoanalytic Study of the Child,* Vol. 39, 1958)
h. Kestenberg, Judith S.
Rhythm and Organization in Obsessive-Compulsive Development (Reprinted from *The International Journal of Psycho-Analysis,* Vol. 47, 1966)

Recognition of Facial Expression
a. Jenness, Arthur
The Recognition of Facial Expressions of Emotion (Reprinted from *Psychological Bulletin,* Vol. 29, 1932)
b. Frois-Wittmann, J.
The Judgment of Facial Expression (Reprinted from *Journal of Experimental Psychology,* Vol. 13, 1930)
c. Landis, Carney
The Interpretation of Facial Expression In Emotion (Reprinted from *Journal of General Psychology,* Vol. 2, 1929)
d. Frijda, Nico H.
The Understanding of Facial Expression of Emotion (Reprinted from *Acta Psychologica,* Vol. 9, 1953)
e. Schlosberg, Harold
Three Dimensions of Emotion (Reprinted from *Psychological Review,* Vol. 61, March 1954)
f. Honkavaara, Sylvia
The Psychology of Expression (Reprinted from *British Journal of Psychology:* Monograph Supplements No. 32, 1961)

Research Approaches to Movement and Personality
a. Eisenberg, Philip
Expressive Movements Related to Feeling of Dominance (Reprinted from *Archives of Psychology,* Vol. 30, No. 211, New York, 1937) May, 1937
b. Bartenieff, Irmgard and Martha Davis
Effort-Shape Analysis of Movement: The Unity of Expression and Function. New York, 1965
c. Takala, Martti
Studies of Psychomotor Personality Tests I. Helsinki, Finland, 1953

Wolff, Charlotte
A Psychology of Gesture. Translated from the French Manuscript by Anne Tennant. 2nd edition. London, 1948